D1452791

Optimal Management of Flow in Groundwater Systems

OPTIMAL

MANAGEMENT

OF FLOW IN

GROUNDWATER

SYSTEMS

DAVID P. AHLFELD

ANN E. MULLIGAN

Department of Civil and Environmental Engineering
University of Massachusetts, Amherst

ACADEMIC PRESS

A Harcourt Science and Technology Company

San Diego   San Francisco   New York   Boston   London   Sydney   Tokyo

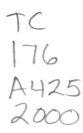

This book is printed on acid-free paper. ∞

Academic Press
A Harcourt Science and Technology Company
525 B Street, Suite 1900, San Diego, California 92101-4495, USA
http://www.apnet.com

Academic Press
24-28 Oval Road, London NW1 7DX, UK
http://www.hbuk.co.uk/ap/

Library of Congress Catalog Card Number: 99-66036

International Standard Book Number: 0-12-044830-0
International Standard Book Number: 0-12-044831-9 (CD-ROM)

PRINTED IN THE UNITED STATES OF AMERICA
99 00 01 02 03 04 EB 9 8 7 6 5 4 3 2 1

CONTENTS

1

GROUNDWATER FLOW MANAGEMENT

2

GROUNDWATER SIMULATION

3

BUILDING THE MANAGEMENT FORMULATION

4

SOLVING THE MANAGEMENT FORMULATION

5

USING THE MANAGEMENT MODEL

6

ADVANCED LINEAR FORMULATIONS

7

FORMULATIONS WITH BINARY VARIABLES

8

FORMULATIONS WITH NONLINEAR FUNCTIONS

PREFACE

When the well's dry, we know the worth of water.
—Benjamin Franklin, 1746

Historically, groundwater has been a reliable, clean, and virtually un-limited water supply for much of the world population. Imagine, you dig a hole in the ground, just about anywhere, and sooner or later you come upon water! What a marvelous resource. Recent pressures on groundwa-ter resources have diminished the ready availability of potable groundwater in many parts of the world. The growth of the human population and in-creased consumption on a per capita basis have depleted the available sup-plies of groundwater, especially in arid regions. In many developed parts of the world the quality of water has been affected by the widespread use of synthetic industrial and agricultural chemicals and by saltwater intrusion in coastal regions. At the same time, awareness of the importance of uncon-taminated groundwater for maintaining ecosystems in wetlands and streams has increased.

As demands on groundwater have increased, the need for careful man-agement of this resource has become apparent. Quantitative techniques are required to best satisfy the competing demands for consumptive use, ecological protection, and maintenance of water quality. Management of these complex systems often benefits from mathematical simulation mod-els. The addition of optimization methods results in a powerful set of tools with which to maximize utilization of water, minimize adverse impacts on the environment, or minimize the costs of achieving management objec-tives. We have written this book to provide access to these tools for a wide audience.

The intention of this book is to provide a complete introduction to the science and art of combined simulation–optimization modeling of ground-water flow. The science introduced in this book focuses on the mathematical coupling of simulation models with optimization methods. A variety of mathematical and numerical issues are explored in detail. The art in management modeling involves the creation of optimization formulations that translate policy and design objectives into accurate mathematical criteria imposed on the simulation model. We convey this skill through numerous examples and descriptions of the implications of alternative model building choices.

We have designed the book to be accessible to a variety of readers. In early chapters, basic concepts are introduced for simple forms of the simulation and management models. In subsequent chapters, successive levels of complexity are added as additional capabilities are introduced. We presume that the reader is comfortable with linear algebra and differential calculus. Our intention is that this book be useful to the experienced groundwater simulation modeler, who may find this book a means of rapid entrance to this set of techniques; to students of groundwater modeling, hydrogeology, or water resources management, who may find this a useful course text or course supplement; and to those involved in performing advanced research in this field, who may find this a useful reference book.

We are indebted to our many colleagues who have guided the development of our understanding of this fascinating field and advised in the writing of this book. Special acknowledgments go to Steven Gorelick, George Pinder, Robert Willis, and William Yeh. We are grateful to Robert Marryott, Eric Reichard, R. Guy Riefler, Charles Sawyer, and Kathleen Yager, who reviewed portions of this book. Finally, we offer thanks to our spouses, Victoria Dickson and Greg Hirth, for their support and encouragement.

David Ahlfeld
Ann Mulligan

LIST OF VARIABLES

C_j	objective function coefficient that includes all terms independent of pumping rate q_i	6
C_0	constant flux from stream to aquifer per unit river stage	6,8
$\mathbf{d}_k$	search direction at iteration k	8
$\mathbf{e}_i$	vector of zeros with a one in the ith element	5
$f()$	general function	1,3,4,5,6,7,8
f_i	set of stress facilities with overlapping numerical grid cells	7
F_i	specified flux that must be maintained at location i	3
F	cumulative probability distribution	6
$g()$	general constraint function	1
g	gravitational constant	8
h_i	head at location i	1,3,4,7
h_i^l, h_i^u	lower and upper bounds on head at location i	1,3,4,8
h_{ik}^d	constraint on difference between heads at locations i and k	1
h_k^d	specified bound on head differences for constraint k	3,4,8
h_i^0	initial head at location i	3,4
$h_{i,\tau}$	head at location i during management period τ	6
h_L	head loss	8
H_i	head at location i in the adjacent water body	3
H	total head needed to lift water (elevation + head loss)	8
H	specified reference head	8
i	interest rate	6,7
J_Λ	Jacobian matrix of the constraint equations	6
K_x, K_y, K_z	hydraulic conductivity in coordinate directions x, y, z	2
K	hydraulic conductivity	3,6,8
K_f	friction loss term	8
l	number of head constraint locations	1,3,4,6,7,8
L	elevation to which water must be lifted	8
m	number of constraints	1,3,4,6,7,8
M	large value in constraints with integer variables	7
n	number of locations considered for application of stress	1,3,4,6,7,8
n_b	number of binary variables	7,8
n_f	number of multiple-cell candidate facilities	7
n_H	number of numerical grid cells in a horizontal stress facility	6

n_w	number of numerical grid cells in a vertical stress facility	6,8
$\mathbf{N}$	matrix of constraint coefficients for nonbasic decision variables	4
N^l, N^u	lower and upper bounds on the number of active stress facilities	7
p_k	specified probability level or reliability measure	6
$\Pr[]$	probability	6
q_j	pumping stress at location j	1,3,4,6,7
q_j^l, q_j^u	lower and upper bound on stress q_j	1,3,7
$q_{j,t}$	stress rate at location j during management period t	1,6
q_j^{ru}, q_j^{eu}	upper bounds on recharge and extraction for stress q_j when direction of stress is unspecified	6
q_j^e, q_j^r	extraction and recharge components of stress q_j	6
$\mathbf{q}$	vector of pumping rates	4,5,6
$\mathbf{q}_0$	vector of initial pumping rates with elements q_j^0	4,6
$\mathbf{q}_{\Delta j}$	vector of pumping rates that differs from $\mathbf{q}_0$ only in the jth element by an amount $(q_{\Delta j} - q_j^0)$	4,6
$\mathbf{q}^l, \mathbf{q}^u$	vector of lower and upper bounds on stress rates	4
Q	value to constrain the total stress magnitude	1
Q_k^l, Q_k^u	lower and upper bounds on total pumping	3
Q_H	total pumping at a horizontal stress facility	6
Q_w	total pumping at a vertical stress facility	6
R	distance from a well to a reference head H	8
s_j	head drawdown at location j	3,5
S_s	specific storage	2
S	storage coefficient	2
t	management period index for the stress variable	6
T	total number of time periods	1
T_x, T_y	transmissivity in coordinate directions x, y	2
T_m	total number of management periods for the stress variables	6
T_h	total number of management periods for the head variables	6
v_D	Darcy velocity	2
v_p	pore velocity	2

Δq_j	change in stress rate at location j	4
Δx	distance between finite difference nodes in x coordinate direction	2
Δx	distance between head locations k_1 and k_2 in a head difference constraint	3,6
η	porosity	2
θ	angle between velocity component and resultant vector	6
κ_b	objective function coefficient on binary variable X_b	7
λ	vector of dual decision variables	5
Λ_k	constraint equation k	6
μ_k	mean of constraint equation k	6
ρ	density	8
σ	elevation of the base of the aquifer	2
σ_k	standard deviation of constraint equation k	6
τ	managment period index for head variables	6
ϕ_1, ϕ_2, ϕ_3	specified functions that define boundary conditions	2
φ	specified constant	3
ω	weighting function	5,8

1

GROUNDWATER
FLOW MANAGEMENT

1.1 INTRODUCTION

1.1.1 THE GROUNDWATER RESOURCE

Demand for fresh water will increase as the world's population continues to grow. Although water is abundant on earth, fresh water accounts for
only about 2.5% of global water reserves. Of this fresh water, approximately
30% is stored as groundwater and 0.3% as rivers and lakes, while the remaining reserves are inaccessibly held in glaciers, ice caps, soil moisture,
and atmospheric water vapor. Global water consumption during 1996 has
been estimated to be more than 50% of available fresh water. Meanwhile,
water use estimates for 2025 project that over 70% of accessible fresh water will be consumed and that 35% of the world's population will encounter
chronic water shortages.

Groundwater has traditionally provided an important source of potable
water because it is readily available in many locations and requires little
or no treatment. Groundwater accounted for approximately 22% of water
use in the United States in 1985 and accounts for 10–50% of water use
in most European countries. In the United States, 56% of the population
uses groundwater as their source of drinking water and in some states, for
example Florida and New Mexico, as much as 90% of potable water is
supplied from groundwater.

In some regions, where demand for groundwater is highest, the continued availability of groundwater may be limited. Overpumping from the
Ogallala aquifer in the midwestern United States has resulted in the aver-

age saturated thickness declining from approximately 17.5 meters in 1930 to 2.5 meters in 1983. Continued groundwater mining may also be precluded by the adverse side effects of overpumping. In many locations of the United States, excessive groundwater withdrawals have led to ground surface subsidence. The Houston–Galveston and Tucson areas have seen subsidence of approximately 3 meters and parts of the San Joaquin Valley, California, have experienced about 9 meters of subsidence. Overpumping also affects wetlands and other surface water ecological systems by interfering in the natural hydrologic cycle.

Stresses on groundwater resources can also lead to groundwater quality degradation. Large-scale groundwater contamination has occurred as a result of excessive use of agricultural chemicals. Overpumping along the eastern seaboard has resulted in saltwater intrusion in many coastal regions, rendering portions of these aquifers unusable for human consumption. Water quality has also been degraded as a result of localized industrial pollution. In 1988, 72% of proposed and final sites on the U.S. National Priorities List, which are regulated and undergo cleanup under the Superfund program, had some form of groundwater contamination.

The future portends increasing demands for groundwater. Continued utilization of this important resource will require consideration of physical, human health, and ecological factors. The analysis and management of groundwater resources will require the coordinated efforts of hydrogeologists, hydraulic engineers, and water resource planners and managers. These professionals can benefit from sophisticated computational tools to analyze these complex systems.

1.1.2 MANAGEMENT OF GROUNDWATER RESOURCES

Groundwater simulation models are now commonly used for analysis and decision making in a wide variety of groundwater-related problems. These models are capable of simulating groundwater flow, hydraulic heads, and transport of contaminants in groundwater and are now in routine use for problems including water supply management, pollutant control, and ecosystem protection. In many cases, these models are used to predict the response of the groundwater system to human-induced modifications.

The relationship between the simulation model and model inputs and outputs can be depicted as shown in Figure 1.1, where the inputs to the simulation model are divided into those describing physical parameters and numerical characteristics and those describing management controls. The outputs of the simulation model describe the state of the system that results from the inputs. The management problem can be viewed as determining the appropriate type, location, and settings for controls to produce desired system outputs.

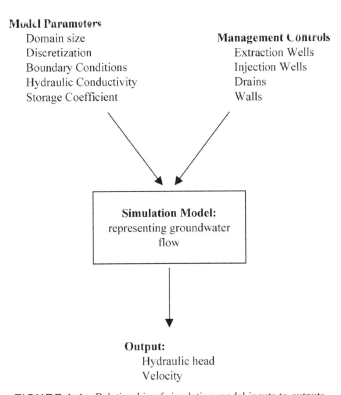

FIGURE 1.1   Relationship of simulation model inputs to outputs.

As the complexity of a problem increases, the ability of the decision maker to determine intuitively the set of controls that will yield a satisfactory solution diminishes. This book provides an additional set of analytical tools to assist the analyst in solving complex management problems. These tools are based on coupling groundwater simulation modeling with methods of optimization. Although it is possible to couple any groundwater simulation model with optimization, this book focuses on simulation models of groundwater that predict hydraulic head and velocity in general groundwater systems. The groundwater management problem can then be stated in the form of an optimization problem that can be solved mathematically.

1.1.3 ORGANIZATION OF THIS BOOK

This book presumes that the reader is generally familiar with groundwater simulation models and the fundamental principles of groundwater flow and transport. The book contains a brief chapter reviewing groundwater models and numerical methods to provide necessary background for using these techniques in the context of optimization. Optimization methods are introduced as needed to solve specific classes of formulations. The first chapter describes the groundwater flow management problem in general terms and provides several examples of potential control problems and formulations. The groundwater simulation model and numerical methods are reviewed in the second chapter. Chapter 3 provides details on constructing a linear management formulation for confined steady groundwater problems with continuous decision variables. Chapter 4 describes how the groundwater flow management formulation is solved using the simplex algorithm. Chapter 5 describes the wealth of information available from the solution of the linear formulation. In Chapters 6, 7, and 8 additional features that can be incorporated into the basic formulation are described, including multiple time period, integer, and nonlinear problems. At the end of each chapter is a section providing selected references and brief histories of the methodologies presented in the book.

1.2 OPTIMIZATION APPROACH TO GROUNDWATER MANAGEMENT

Optimization methods have been used for decision making for many years. Applications of optimization techniques abound in fields such as business, engineering, the physical sciences, and the social sciences. To solve a problem using optimization methods, components of the problem must be assembled into a mathematical structure. An optimization problem has three key elements: the objective function, the constraints, and the decision variables. Two types of optimization formulations can be constructed with these elements: unconstrained problems, which include an objective function and decision variables, and constrained problems, which contain all three elements. The optimization formulation is a model of the management decision-making process and is referred to as a management model.

The first element of the optimization formulation is the set of decision variables. For design problems, the decision variables describe the controls that are to be designed. The values taken by these variables define the solution of the problem.

The second component of a management model is the set of constraints, which impose restrictions on the values that can be taken by the decision variables. For example, decision variables may be required to be continuous or integer. In addition, upper and lower bounds may be imposed on the value that decision variables may take. Constraint functions involving multiple decision variables may also be defined. Bounds on the values of these constraint functions may be imposed.

Optimization methods are generally used to solve problems in which multiple solutions satisfy all of the constraints. The task is to identify the solution that is "best" by some appropriate measure. This measure takes the form of an objective function, which is stated in terms of the decision variables. This function may be maximized or minimized depending upon the application.

The optimization approach requires formal definition of the decision variables, the constraints to be imposed on the management model, and the objective to be optimized. The objective and constraints are translated into mathematical functions of the decision variables to produce the optimization formulation. The formulation that results is then solved using one of a variety of optimization algorithms. The solution to a well-posed optimization formulation consists of values for the decision variables that optimize the objective function while satisfying all constraints on decision variable values.

1.2.1 FEATURES OF THE OPTIMIZATION APPROACH

A key feature of the optimization approach to groundwater management is the direct coupling between the simulation model and the management model. This linking between models is accomplished by formulating the constraints and objectives as direct functions of the management control inputs and simulation model outputs. For much of the work described here, the management controls or decision variables will be the location and magnitude of stresses imposed on the groundwater system. Stress should be understood to mean a flow imposed on the aquifer at a specified location, over a specified duration, and at a specified rate. The stress may result in the extraction of water from or recharge of water to the aquifer. These imposed stresses are applied through facilities such as wells, drains, infiltration basins, or interception structures. The simulation model output will be the hydraulic heads and other properties of the system such as velocities.

The optimization approach can be viewed in contrast to the conventional iterative approach to design, which is depicted in Figure 1.2. Here, a simulation model is constructed to provide predictions of groundwater flow that result from alternative sets of management controls. A specific set of controls is selected and simulated. The model output is compared with specified design criteria. The process is repeated until a satisfactory result is determined or analysis resources (e.g., available time or budget) are exhausted.

The optimization approach is depicted in Figure 1.3. Again, a simulation model is required. However, here, the design criteria are used to define the optimization formulation. The simulation model provides the feedback necessary to evaluate whether specific control options meet the design criteria. The resulting formulation is directly solved to determine the optimal set of management controls.

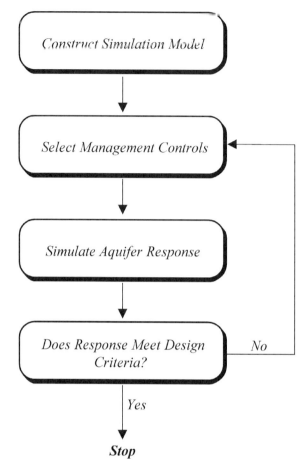

FIGURE 1.2   Iterative approach to design of groundwater controls.

The optimization approach has numerous benefits. At a mechanistic level it can be viewed as effectively automating the trial-and-error process and efficiently searching all possible solutions to a particular problem. In some cases, for a particular set of constraints, no feasible solution will exist. This circumstance will be automatically and conclusively identified by the optimization algorithm. From an analysis perspective, the optimization approach is an invaluable aid to the analyst because it requires that decision makers and relevant stakeholders quantify goals and objectives. From a policy perspective, the optimization formulation can be viewed as a model of the design process. The parameters of the optimization formulation can be manipulated in a sensitivity analysis to determine the relationships between constraints imposed on the problem and objectives to be optimized.

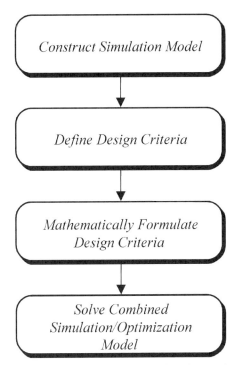

FIGURE 1.3 Optimization approach to design of groundwater controls.

Before proceeding, we introduce a disclaimer with regard to the use of the word optimization. In this book, formal optimization methods are used to solve the problem at hand. Algorithms will identify the optimal solution to a problem defined by a given objective function and set of constraints. However, the use of mathematical optimization does not guarantee the best solution. The results of the optimization depend on the reliability of the underlying simulation model and on the form of the objective function and constraints. The optimization approach is best viewed as providing an additional tool that can help the perceptive analyst understand the system under study.

1.2.2 GENERAL CONTINUOUS VARIABLE FORMULATION

We begin by defining the relationship between simulation model input and output as

$$h_i = h_i(q_1, q_2, \ldots, q_n) \tag{1.1}$$

where h_i is the head at location i and each q represents one of the n stresses imposed on the groundwater system. These stresses are flow rates with dimension of volume per time applied at specified locations such as wells, drainage lines, infiltration basins, or interception structures. A positive value of stress implies extraction from the aquifer; a negative value implies recharge. The relationship given by (1.1) is determined by a simulation model of the groundwater flow system.

A general continuous-variable constraint on model input and output takes the form of a requirement on a function of stresses and heads

$$g(q_1, q_2, \ldots, q_n, h_1, h_2, \ldots, h_l) \leq 0 \tag{1.2}$$

where l different heads may be involved.

Typical constraint forms are simple upper and lower bounds on heads

$$h_i^l \leq h_i \leq h_i^u \tag{1.3}$$

and constraints on the differences in head values

$$h_i - h_k = h_{ik}^d \tag{1.4}$$

where h_{ik}^d is the specified difference in head between locations i and k.

Typical constraint forms on stresses include simple bounds on stress at specified locations

$$q_j^l \leq q_j \leq q_j^u \tag{1.5}$$

and requirements on total stress magnitude

$$\sum_{j=1}^{n} q_j \geq Q \tag{1.6}$$

The objective function is a general function of both the stresses and the model outputs and takes the form

$$f(q_1, q_2, \ldots, q_n, h_1, h_2, \ldots, h_l) \tag{1.7}$$

Typical objective functions seek to optimize the stress imposed on the aquifer to, say, minimize or maximize the total stress,

$$f = \sum_{j=1}^{n} q_j \tag{1.8}$$

The general formulation for continuous-variable problems takes the form

$$\text{minimize } f(q_1, q_2, \ldots, q_n, h_1, h_2, \ldots, h_l) \tag{1.9}$$

$$\text{such that}$$

$$g_k(q_1, q_2, \ldots, q_n, h_1, h_2, \ldots, h_l) \leq 0, \qquad k = 1, \ldots, m \tag{1.10}$$

Because the heads depend on the stresses, this formulation could be defined as solely a function of the stresses. However, for clarity of presentation we will retain the explicit presence of the heads. This formulation will be amended for the special case of integer variables in Chapter 7. Note that this formulation encompasses both minimization and maximization because maximizing the negative of a function is equivalent to minimizing the original function.

The optimization approach consists of determining the specific set of objectives and constraints that address the problem under study, formulating these into a set of functions that fit the general form shown here, and solving the resulting optimization problem.

1.3 APPLICATIONS OF THE OPTIMIZATION APPROACH

The optimization approach can be applied to a wide variety of problems. In this section, the use of optimization is demonstrated in several typical applications. These simple examples are intended to provide an indication of the breadth of problems that can be solved. The formulations presented are only a few of many possible ways in which these problems can be stated.

1.3.1 POLLUTANT CONTROL

Consider a portion of an aquifer that must be isolated from the rest of the aquifer. This requirement might result from the existence of a body of groundwater of degraded quality arising from a pollutant source. The pollutant control problem consists of producing a hydraulic regime that isolates a specified portion of an aquifer. In Figure 1.4, a plume of polluted groundwater emanating from a single pollutant source is depicted in a vertically exaggerated cross section. The design criteria are that the direction of groundwater flow along the perimeter of the plume must be reversed so that expansion of the plume is halted.

Criteria on the direction of groundwater flow can be formulated as constraints on the difference in hydraulic head at selected locations. Heads that satisfy such constraints will produce velocities that have a positive component in the direction suggested by the location of the selected head points. In the figure, four pairs of locations are identified. The objective of the problem can be formulated to minimize the total cost of stress required to produce inward flow around the volume to be contained. In the figure, five wells, with different locations and vertical screen placements, are specified from which the optimization algorithm will select appropriate stress rates.

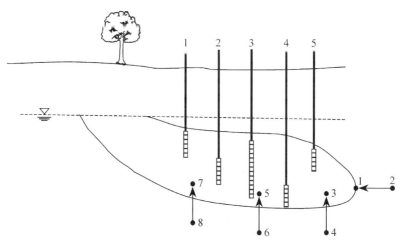

FIGURE 1.4 Pollutant control example: The stress rates from five wells must be determined so that the constrained differences in hydraulic head at the four locations shown are met.

The preceding qualitative problem description can be stated in the following optimization formulation:

$$\text{minimize } f = \sum_{j=1}^{5} \alpha_j q_j \tag{1.11}$$

such that

$$h_2 - h_1 \geq h_1^d \tag{1.12}$$

$$h_4 - h_3 \geq h_2^d \tag{1.13}$$

$$h_6 - h_5 \geq h_3^d \tag{1.14}$$

$$h_8 - h_7 \geq h_4^d \tag{1.15}$$

where α_j is the cost per unit stress at well j and h_i^d is the required head difference for the ith head pair.

1.3.2 MINING AND CONSTRUCTION DEWATERING

Subsurface construction often involves excavation beneath the water table. Such activities might include foundation excavation, mining, and tunnel construction. It may be important both to remove water from the excavated space and to reduce the pore pressure in adjoining soils to enhance stability and avoid collapse. In these cases the design criteria may be to require that the hydraulic heads be maintained beneath the base of the excavation and that a scheme be found to minimize the number of wells or the amount of stress required to achieve these criteria.

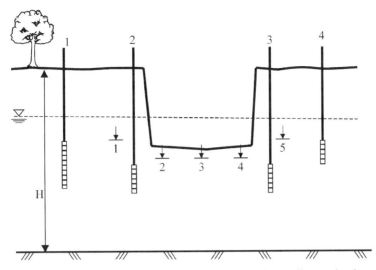

FIGURE 1.5 Dewatering example: The stress rates from four wells must be determined so that hydraulic heads at the five locations indicated are below the specified elevations.

An example of this problem is depicted in Figure 1.5, where the initial head is depicted by the dashed line. The head in the aquifer is to be maintained below specified levels at the five locations indicated by the points of the arrows. This lowering of the water table is to be accomplished by pumping from any combination of the four wells. The objective is intended to minimize the total rate at which water is pumped. The optimization formulation can be stated as

$$\text{minimize } f = \sum_{j=1}^{4} q_j \tag{1.16}$$

such that

$$h_1 \leq h_1^u \tag{1.17}$$

$$h_2 \leq h_2^u \tag{1.18}$$

$$h_3 \leq h_3^u \tag{1.19}$$

$$h_4 \leq h_4^u \tag{1.20}$$

$$h_5 \leq h_5^u \tag{1.21}$$

where h_i^u is the upper bound on head at location i.

For short-term dewatering projects, the construction cost of the dewatering system may exceed the operation cost. The formulation posed here accounts for only operating costs. The introduction of construction costs into the formulation will be discussed in Chapter 7.

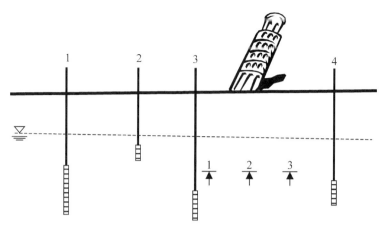

FIGURE 1.6 Subsidence control: Extraction from the four wells must not reduce the hydraulic head in the aquifer below the elevations shown.

1.3.3 SUBSIDENCE CONTROL

In some aquifers excessive drawdown causes subsidence of the overlying geologic strata. Water supply managers may need to distribute pumping in space and time so as to prevent excessive drawdown in critical portions of the aquifer. An example of this situation is depicted in Figure 1.6, where four wells are available for pumping. The hydraulic head under nonpumping conditions is depicted by the dashed line. During pumping, head must be maintained above a specified level at the three locations indicated by the points of the arrows beneath the leaning tower.

Assuming that stress can be varied over time, a formulation can be posed in which total withdrawals are maximized subject to requirements on hydraulic head at critical locations. This formulation takes the form

$$\text{maximize } f = \sum_{j=1}^{4} \sum_{t=1}^{T} q_{j,t} \tag{1.22}$$

such that

$$h_{1,t} \geq h^l \,, \qquad t = 1, \ldots, T \tag{1.23}$$

$$h_{2,t} \geq h^l \,, \qquad t = 1, \ldots, T \tag{1.24}$$

$$h_{3,t} \geq h^l \,, \qquad t = 1, \ldots, T \tag{1.25}$$

where $q_{j,t}$ is the stress at location j in time period t, T is the total number of time periods, and the constraints on head are imposed at every time step.

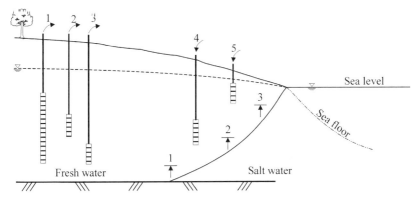

FIGURE 1.7 Saltwater intrusion control: Three extraction wells, two recharge wells, and three lower bound head constraints are used to meet water supply demands and control saltwater intrusion.

An alternative formulation would require that water demand be satisfied in each time period, while producing the smallest amount of subsidence. This can be accomplished by maximizing the heads such that the summation of stresses is greater than or equal to the stress demand in period t, Q_t.

$$\text{maximize } f = \sum_{i=1}^{3} \sum_{t=1}^{T} h_{i,t} \tag{1.26}$$

such that

$$\sum_{j=1}^{4} q_{j,t} \geq Q_t , \qquad t = 1, \ldots, T \tag{1.27}$$

1.3.4 SALTWATER INTRUSION

In many coastal regions of the world, excessive groundwater pumping produces saltwater intrusion into freshwater aquifers. Water managers may require pumping strategies that halt or reverse the intrusion of saltwater while continuing to meet demands for fresh water. This problem has similarities to the pollutant control problem and can be formulated in a similar fashion. Here, a more complex formulation is proposed that involves both extraction for water supply and injection to control saltwater intrusion. An example is depicted in Figure 1.7, where a saltwater wedge is intruding into a freshwater aquifer. The control of intrusion is ensured by requiring that freshwater heads at the three locations indicated by the points of the arrows be maintained at levels sufficiently high, relative to sea level, that intrusion is prevented. Extraction can be carried out from any combination of the three locations shown. Water can be injected at either of the two locations shown.

One possible formulation for this problem maximizes net withdrawals while ensuring that saltwater intrusion is halted. It is assumed that all injected water is derived from water extracted upgradient. Constraints are also imposed on the amount of stress possible at each location. The resulting formulation is

$$\text{maximize } f = \sum_{j=1}^{5} q_j \tag{1.28}$$

such that

$$h_1 \geq h_1^l \tag{1.29}$$

$$h_2 \geq h_2^l \tag{1.30}$$

$$h_3 \geq h_3^l \tag{1.31}$$

$$0 \leq q_j \leq q_j^u, \quad j = 1, \ldots, 3 \tag{1.32}$$

$$q_j \leq 0, \qquad\qquad j = 4, 5 \tag{1.33}$$

where h_i^l is the lower bound on head at location i. Note that the first three stresses are required to be nonnegative while the second two are required to be nonpositive, implying extraction and recharge, respectively.

1.3.5 WETLAND PROTECTION FROM DEWATERING

Wetlands play an important role in maintaining regional ecological systems. In settings where wetlands are hydraulically connected to groundwater, the implementation of large-scale groundwater extraction can interfere with wetland hydrology through interception of discharging waters or drainage of wetland waters. The use of groundwater flow management formulations in such a setting can be demonstrated by considering the example depicted in Figure 1.8, where extraction is considered at some combination of the three wells indicated. The objective is to maximize the extraction of water while controlling the impact on the wetland hydrology.

This problem could be formulated in a manner similar to the saltwater intrusion problem where heads near the wetland are required to be maintained above a specified level. Alternately, the problem could be formulated to ensure that the total groundwater flux to the wetland is maintained.

Assume that the total groundwater flow to the wetland is known and is given by Q_0. Acknowledging that pumping will produce some interception of groundwater discharging to the wetland, it is required that the total flow in the presence of pumping be greater than a fraction, β, of Q_0. The

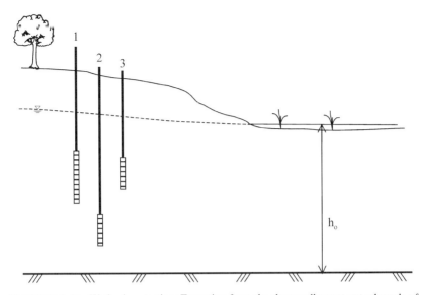

FIGURE 1.8 Wetland protection: Extraction from the three wells must not adversely affect the adjacent wetland.

formulation is written to maximize extraction as

$$\text{maximize } f = \sum_{j=1}^{3} q_j \tag{1.34}$$

such that

$$\sum_{i=1}^{m} C_i (h_i - h_o) \geq \beta Q_o \tag{1.35}$$

where a set of m locations, indexed by i, are selected in the aquifer near the aquifer–wetland interface, h_i is the head at location i, h_o is the head in the wetland, and C_i is the conductance between location i and the wetland.

This formulation presumes steady operation and no seasonal variation in wetland head. A more complicated formulation could be posed to select pumping rates in each season with constraints on flow to the wetland imposed only in seasons in which dry conditions prevail.

1.4 NOTES AND REFERENCES

The global water use estimates presented in Section 1.1 are from Hinrichsen *et al.* (1997). Global water resources are reviewed by Gleick (1993) and freshwater resources in the United States are described in Francko and

Wetzel (1983) and Thompson (1999). The encyclopedia by van der Leeden *et al.* (1990) summarizes U.S. and global water quantity and quality characteristics.

The application of optimization techniques to groundwater flow began in 1970 with the paper by Deninger. Prior to this, optimization had been combined with porous media models in oil reservoir applications, with the first work on the topic by Lee and Aronofsky (1958). Both of these works relied upon analytical solutions to the relevant governing equations. The first combination of numerical simulation models with optimization for groundwater applications was by Maddock (1972). Since these early works, the application of optimization to groundwater problems has grown substantially.

The examples presented in Section 1.3 draw from a variety of techniques developed by researchers over many years. References to these techniques, formulations, and applications are provided in the notes for Chapters 3–8. Introductions to the field and comprehensive reviews of the literature related to the optimization–simulation approach are presented by Gorelick (1983), Yeh (1992), Ahlfeld and Heidari (1994), and Wagner (1995). Several books covering management modeling for water resources and groundwater systems are also available. Surface water and groundwater issues are covered by Mays and Tung (1992) and Mays (1997), and groundwater problems are addressed by Willis and Yeh (1987) and Gorelick *et al.* (1993). Researchers in the former Soviet Republics have also been active in developing optimization-based models for groundwater management (see, for example, Veselov *et al.*, 1992).

The scope of this book is limited to management models that combine optimization with simulation models for saturated groundwater flow. Significant progress has been made in applying optimization to other aspects of subsurface phenomena. The interested reader may wish to pursue extensions of the methods described in this book using the selected references that follow. More thorough references can be found in the books just cited. Formulations that couple optimization with subsurface solute transport simulation models have been devised for groundwater remediation design. Processes included are steady pumping (Gorelick *et al.*, 1984), time-varying pumping (Ahlfeld, 1990; Culver and Shoemaker, 1992), nonequilibrium adsorption (Haggerty and Gorelick, 1994), and biologically driven solute degradation (Minsker and Shoemaker, 1996). Simulation model parameter uncertainty has been incorporated by Wagner and Gorelick (1987). Optimization has also been applied to soil vapor extraction in the unsaturated zone (Sun *et al.*, 1996).

2

GROUNDWATER
SIMULATION

This chapter provides a brief introduction to some of the numerical methods and associated notation that will be used in the remainder of the book. This chapter presumes that the reader has limited familiarity with numerical methods as applied to groundwater flow problems.

Groundwater management models require a reliable and accurate groundwater simulation model. Modern groundwater simulation models are capable of solving heterogeneous, transient, nonlinear problems on complex domains. Efficient numerical methods have been developed and reliable computer codes are widely available. In this chapter we review the common equations used to analyze groundwater flow, present key elements of the finite difference method, and discuss the numerical error associated with implementing a numerical solution.

2.1 GOVERNING EQUATIONS FOR GROUNDWATER FLOW

The governing equation for groundwater flow in three dimensions is written as

$$\frac{\partial}{\partial x}\left(K_x \frac{\partial h}{\partial x}\right) + \frac{\partial}{\partial y}\left(K_y \frac{\partial h}{\partial y}\right) + \frac{\partial}{\partial z}\left(K_z \frac{\partial h}{\partial z}\right) + q = S_s \frac{\partial h}{\partial t} \qquad (2.1)$$

where the subscripted K terms are the hydraulic conductivity in each coordinate direction, q is the source–sink term representing flow rate per unit volume, and S_s is the specific storage. This form of the equation presumes that the coordinates are aligned with the principal directions of anisotropy. In most field applications these equations are imposed on a finite domain with boundary conditions of the form

$$\phi_1 h + \phi_2 \frac{\partial h}{\partial n} = \phi_3 \qquad (2.2)$$

where ϕ_1, ϕ_2, and ϕ_3 are specified functions of space and time. For a given location and time there are three main types of boundary conditions that can be generated using (2.2). A Dirichlet, or first type, condition results when $\phi_2 = 0$, ϕ_1 is a nonzero real number, and ϕ_3 is any real number. For groundwater problems, this is also called a constant head condition. A constant flux boundary arises in the creation of a Neumann, or second type, condition where $\phi_1 = 0$, ϕ_2 is nonzero, and ϕ_3 is any real number. Finally, when ϕ_1, ϕ_2, and ϕ_3 are all nonzero, then the Robbins, or mixed, condition arises. The first, second, and third type conditions described are linear boundary conditions. Nonlinear boundary conditions are possible when, for example, the coefficients ϕ_1, ϕ_2, and ϕ_3 depend upon h.

The groundwater flow equation is often integrated in the z coordinate direction to yield, in two spatial dimensions,

$$\frac{\partial}{\partial x}\left(T_x \frac{\partial h}{\partial x}\right) + \frac{\partial}{\partial y}\left(T_y \frac{\partial h}{\partial y}\right) + Q = S \frac{\partial h}{\partial t} \qquad (2.3)$$

where the subscripted T terms are the transmissivity in each coordinate direction, Q is the vertically averaged source–sink term, and S is the storage coefficient. This form of the equation assumes that the aquifer is confined.

If the aquifer is unconfined and the coordinate system described by (2.1) is oriented so that z is aligned with the gravitational vector (Figure 2.1), then the two-dimensional form of the flow equation becomes

$$\frac{\partial}{\partial x}\left(K_x(h - \sigma)\frac{\partial h}{\partial x}\right) + \frac{\partial}{\partial y}\left(K_y(h - \sigma)\frac{\partial h}{\partial y}\right) + Q = \eta_d \frac{\partial h}{\partial t} \qquad (2.4)$$

where σ is the elevation of the base of the aquifer so that $(h - \sigma)$ gives the aquifer thickness and η_d is the drainage porosity. Note that this unconfined equation is nonlinear in the state variable h.

Solution of the full three-dimensional unconfined case is complicated because the elevation of the free surface of the domain is unknown. This problem requires incorporating a complex nonlinear boundary condition on the free surface. As an alternative, the quasi-three-dimensional form of the flow equation is widely used in which the vertical dimension is subdivided into a series of layers. The head in each layer is modeled as two-dimensional horizontal flow with the connection between layers represented by a conductance term, which relates the vertical flow between layers to the head difference between those layers.

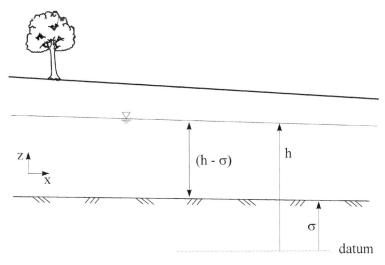

FIGURE 2.1   Variables used in the unconfined aquifer equations.

2.2 DARCY'S LAW

Darcy's law is the fundamental law for describing the relationship between hydraulic gradient and flow. The law is a linear function where the proportionality term is called the hydraulic conductivity. Given the hydraulic head, Darcy's law can be used to calculate the velocity as

$$\mathbf{v}_D = -\mathbf{K}\nabla h \tag{2.5}$$

where $\mathbf{v}_D$ is the velocity vector, $\mathbf{K}$ is the hydraulic conductivity tensor, and ∇h is the gradient of hydraulic head. The velocity $\mathbf{v}_D$, referred to as the Darcy velocity, is actually the flow rate through the porous media per unit cross-sectional area. The true average water velocity is given by the pore velocity. The pore velocity can be determined by dividing the Darcy velocity by the effective porosity of the porous media, η_e,

$$\mathbf{v}_p = \frac{1}{\eta_e}\mathbf{v}_D \tag{2.6}$$

2.3 NUMERICAL METHODS FOR GROUNDWATER FLOW MODELING

Many practical field problems require some form of numerical method for solution because the presence of heterogeneity, irregular domain and

boundary conditions, or nonlinearities prohibits the use of analytical methods. As a rule of thumb, if a problem is sufficiently simple to be addressed
by analytical methods, its analysis is unlikely to benefit from the use of optimization. Hence, we focus here on problems solved by numerical methods.

The widespread growth of the use of computers by practitioners has been
accompanied by the development of a wide variety of numerical methods
for solving groundwater flow problems. These solution techniques include
the finite difference method, the finite element method, analytical elements,
integrated finite differences, and the boundary integral method.

It is important for the reader to understand some basic concepts from
numerical methods as many of the procedures discussed in this book build
on these ideas. Without lending preeminence to any particular method, we
shall describe the finite difference method applied to the groundwater flow
equation. This method has the advantages of being simple and widely used
in some of the most popular simulation models such as the U.S. Geological
Survey code MODFLOW.

2.3.1 FINITE DIFFERENCE METHOD

The basic approach of most numerical methods for the solution of partial
differential equations is to convert the differential equation into a system
of algebraic equations defining the state variable at discrete node points. A
key step in constructing a finite difference solution is the discretization of
the domain. A primary component of this discretization is the node or cell,
which represents a specific location in space and time. Solution of the finite
difference equations implies determining the value of the state variable at
each node. For convenience, the nodes may be identified by an indexing
scheme. Example discretizations with indexing schemes are indicated in
Figures 2.2 through 2.5. There are four possible levels of discretization,
namely three spatial dimensions and one temporal dimension. Therefore,
the highest indexing order is for the node i, j, k, t, where i, j, and k are
spatial indices and t is the time index. The solution of the problem implies
determining the value of the state variable throughout the domain. The
numerical method determines the value at each node, and therefore the
state variable is indexed to a node as h_i, which implies the head value at
location i.

The basic concept of the finite difference method is to approximate
derivatives using the difference in values at adjoining nodes. The first
derivative can be approximated by forward differencing

$$\frac{dh_i}{dx} = \frac{(h_{i+1} - h_i)}{(x_{i+1} - x_i)} = \frac{(h_{i+1} - h_i)}{\Delta x} \tag{2.7}$$

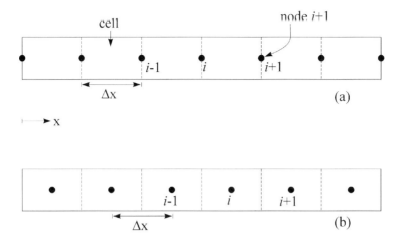

FIGURE 2.2   Grid discretization in one dimension. In (a), the nodes are at the edge of each grid cell. Part (b) depicts a block-centered discretization.

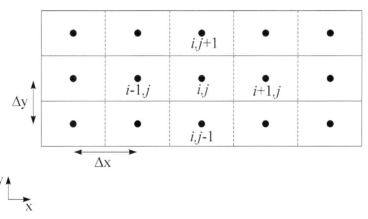

FIGURE 2.3   Two-dimensional block-centered discretization.

Another first-derivative approximation can be made using a backward differencing scheme

$$\frac{dh_i}{dx} = \frac{(h_i - h_{i-1})}{(x_i - x_{i-1})} = \frac{(h_i - h_{i-1})}{\Delta x} \tag{2.8}$$

The second derivative can be approximated by

$$\frac{d^2h}{dx^2} = \frac{h_{i+1} - 2h_i + h_{i-1}}{(\Delta x)^2} \tag{2.9}$$

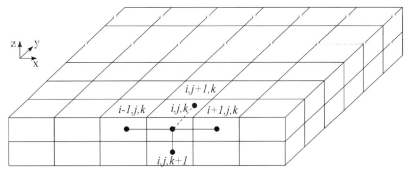

FIGURE 2.4   Three-dimensional discretization.

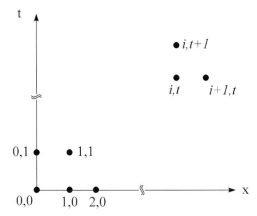

FIGURE 2.5   Temporal and spatial discretization.

EXAMPLE OF FINITE DIFFERENCE APPROXIMATION

The means by which these derivative approximations are used to translate a differential equation into algebraic equations is demonstrated with the following small example. We consider steady one-dimensional flow in a homogeneous media with no sources or sinks. The appropriate governing equation is

$$\frac{\partial}{\partial x}\left(K_x \frac{\partial h}{\partial x}\right) = 0 \tag{2.10}$$

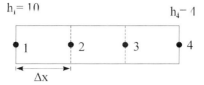

FIGURE 2.6   One-dimensional example discretization.

Because the hydraulic conductivity does not depend on x, it may be extracted from the derivative and eliminated from the equation to form

$$\frac{\partial^2 h}{\partial x^2} = 0 \qquad (2.11)$$

Substituting the second-order approximation, equation (2.9), yields a relationship between any three neighboring nodes.

$$\frac{h_{i+1} - 2h_i + h_{i-1}}{(\Delta x)^2} = 0 \qquad (2.12)$$

Next consider a domain consisting of four nodes as shown in Figure 2.6. The heads at the two end nodes are known to have values of 10 and 4, respectively. Applying (2.12) to nodes 2 and 3 and noting that the spacing Δx is constant yields these two equations:

$$h_3 - 2h_2 + h_1 = 0 \qquad (2.13)$$

$$h_4 - 2h_3 + h_2 = 0$$

These two equations can be rearranged in matrix form as

$$\begin{bmatrix} -2 & 1 \\ 1 & -2 \end{bmatrix} \begin{bmatrix} h_2 \\ h_3 \end{bmatrix} = \begin{bmatrix} -h_1 \\ -h_4 \end{bmatrix} \qquad (2.14)$$

Given the specified values of h_1 and h_4, this system of linear equations can be solved uniquely for the values of the head at the interior node points, which are $h_2 = 8$ and $h_3 = 6$. Note that the head varies linearly, as would be expected because the governing equations and boundary conditions are linear functions and the hydraulic conductivity is constant.

Additional terms in the differential equation can be incorporated by simply adding together the respective finite difference approximations. For two-dimensional transient flow with homogeneous transmissivity, the governing equation is

$$\frac{\partial^2 h}{\partial x^2} + \frac{\partial^2 h}{\partial y^2} = \frac{S}{T} \frac{\partial h}{\partial t} \qquad (2.15)$$

A possible finite difference approximation for this is

$$\frac{h_{i+1,j,t} - 2h_{i,j,t} + h_{i-1,j,t}}{(\Delta x)^2} + \frac{h_{i,j+1,t} - 2h_{i,j,t} + h_{i,j-1,t}}{(\Delta y)^2} = \frac{S}{T} \frac{h_{i,j,t} - h_{i,j,t-1}}{\Delta t}$$

(2.16)

where the nodal values of head are now indexed for both space dimensions and time and a backward-in-time approximation is used for the time derivative. In addition to the boundary conditions required as in the first example, this scheme will require initial conditions. With these in hand, a system of equations can be constructed at time level t given all information at time level $t-1$. At successive time steps, the newly computed solution is used as the initial conditions for the next time level. As with the one-dimensional example, an equation of the form of (2.16) can be written for each node location i, j at time level t. There will be sufficient equations to produce a unique solution at each time level.

It is possible to incorporate many additional features into numerical representations of a groundwater flow system. Methods exist to integrate second and third type boundary conditions, additional source–sink related terms, irregular grid points, and spatial and temporal variability in aquifer properties. The reader should consult the references at the end of this chapter for more information on these techniques. All of the methods have the common feature of translating a differential equation into a system of algebraic equations. Instead of a continuous solution in space, the solution consists of head values at numerous discrete locations.

2.3.2 TAYLOR SERIES

The Taylor series provides the basis for analyzing numerical error in numerical approximation methods and will be used in subsequent chapters to help build the groundwater flow management model. A function that is continuous and infinitely differentiable over an interval can be represented as an infinite series. The general form of this series is called the Taylor series, where the function value at a point $\mathbf{x}$ in the neighborhood of a point $\mathbf{x}_0$ is determined as

$$f(\mathbf{x}) = f(\mathbf{x}_0) + \sum_{i=1}^{n} \left[\frac{\partial f(\mathbf{x}_0)}{\partial x_i} \left(x_i - x_i^0 \right) \right]$$

(2.17)

$$+ \sum_{i=1}^{n} \sum_{j=1}^{n} \left[\frac{\partial^2 f(\mathbf{x}_0)}{\partial x_i \partial x_j} \frac{\left(x_i - x_i^0 \right)\left(x_j - x_j^0 \right)}{2!} \right] + \cdots$$

For one-dimensional functions, the Taylor series in (2.17) can be reduced to

$$f(x) = f(x_0) + \frac{\partial f(x_0)}{\partial x}(x - x_0) + \frac{1}{2!} \frac{\partial^2 f(x_0)}{\partial x^2}(x - x_0)^2 + \cdots$$

(2.18)

The Taylor series is used in many aspects of numerical analysis. For example, the forward difference approximation, shown in (2.7), can be derived from the Taylor series as follows. First, (2.18) is rearranged so that the first-derivative term is isolated on one side of the equation

$$\frac{\partial f(x_0)}{\partial x} = \frac{f(x) - f(x_0)}{(x - x_0)} - \frac{1}{2!} \frac{\partial^2 f(x_0)}{\partial x^2}(x - x_0) - \cdots \tag{2.19}$$

Letting $\Delta x = (x - x_0)$ and substituting into (2.19) results in

$$\frac{\partial f(x_0)}{\partial x} = \frac{f(x) - f(x_0)}{\Delta x} - \frac{1}{2!} \frac{\partial^2 f(x_0)}{\partial x^2}\Delta x - \cdots \tag{2.20}$$

If all terms that include second-order and higher derivatives in (2.20) are eliminated, then the following statement, equivalent to (2.7), can be made:

$$\frac{\partial f(x_0)}{\partial x} \cong \frac{f(x) - f(x_0)}{\Delta x} \tag{2.21}$$

The error in this approximation is the sum of all terms that have been truncated from the infinite series. This error will be dominated by the first truncated term, which is a function of Δx. The forward difference approximation is therefore said to have an error on the order of Δx, also expressed as $O(\Delta x)$.

2.3.3 ACCURACY OF NUMERICAL METHODS

An important consideration when using numerical methods is the error introduced by the approximation to the partial differential equation. Generally speaking, the approximation error increases as the grid becomes coarser. Although grid selection is generally an issue beyond the scope of this book, the issue of numerical accuracy is relevant and important in the solution of optimization problems. As we will see in Chapter 4, the derivatives of head with respect to stress are needed during solution. These derivatives are calculated by perturbation, and both approximation and round-off errors affect the precision of the derivative calculations.

In the previous section, a Taylor series analysis of the forward difference approximation was shown and indicated that the error in the approximation is dominated by the first truncated term in the Taylor series. This truncated term takes the form

$$\frac{1}{2} \frac{\partial^2 f}{\partial x^2}\Delta x \tag{2.22}$$

In this case, we say that the error is proportional to the size of the difference interval. It would seem clear that the best approach would be to make Δx as small as possible, thereby minimizing approximation error. There is a trade-off, however, because several arithmetic computations are performed in solving the system of equations and we must therefore be concerned with round-off error.

Round-off error is produced when the precision of a real number is degraded by successive arithmetic operations. One example of round-off error arises when a difference is taken, using limited precision, between two numbers of nearly identical value. In a typical computer architecture, a single-precision FORTRAN number is stored with only about eight digits of precision. If two numbers, each represented by eight digits of precision with the first six digits of both numbers the same, are subtracted, the resulting difference will be accurate to only two digits of precision.

The implication of the preceding discussion is that truncation error and round-off error are influenced in opposite ways by the numerical step size. Maximum accuracy is achieved when the appropriate balance is made between round-off error and truncation error. Procedures for estimating the two types of errors and for determining an optimal discretization size are presented in references identified at the end of this chapter.

Care must also be taken when considering the numerical time step, where accuracy and solution stability must be taken into consideration. The backward differencing scheme, when applied to the time domain, is unconditionally stable. This implies that the solution will behave in a physically realistic manner regardless of the time step size. Of course, the approximation error is proportional to the step size, so accuracy considerations would advocate a small time step. The forward-in-time differencing scheme, on the other hand, is conditionally stable, meaning that if the time step is larger than some bound, then the solution errors grow unbounded and the solution is unbounded and may be meaningless. This highlights the importance of knowing which numerical scheme is used to solve the groundwater flow equation and understanding the limitations of the method.

2.3.4 LINEAR RESPONSE OF HEAD TO STRESS

The solution methods for management problems described in this book will make extensive use of the principle of superposition. The principle of superposition states that the sum of particular solutions to a homogeneous linear partial differential equation, which is subject to linear boundary conditions of the form (2.2), is also a solution to the differential equation. As stated, the confined flow equation (2.1) is linear but not homogeneous. The equation can be converted to a homogeneous form if the source–sink term q is removed and the impact of sources and sinks is introduced through the boundary conditions. A particular solution to the differential equation is the solution corresponding to a particular set of boundary conditions imposed on the domain. The principle of superposition will allow the impact of combinations of stresses on head to be predicted without resimulation of each new set of boundary conditions.

The principle of superposition forms the basis for the method of images commonly applied in groundwater hydraulics for finding the influence of multiple extraction or recharge wells. Superposition remains valid for transient groundwater flow as long as the governing equation is linear and for boundary conditions (including stresses) that change with time. Superposition is not valid when the governing equation is nonlinear, as is the case for unconfined flow simulation, or when the boundary conditions are nonlinear.

2.4 NOTES AND REFERENCES

The governing equations for groundwater flow introduced in Section 2.1 are presented in many excellent references including Bear (1972, 1979), Freeze and Cherry (1979), Domenico and Schwartz (1997), and de Marsily (1986). Numerical methods for solution of differential equations are described in numerous references including Celia and Gray (1992). Anderson and Woessner (1992) and Huyakorn and Pinder (1983) discuss numerical issues specific to solving groundwater flow problems, including grid design and calibration. The numerical simulation issues and associated effects on optimization procedures discussed in Section 2.3.3 are presented in Gill *et al.* (1981). The basis and limitations of the application of superposition to the groundwater flow equations, described in Section 2.3.4, are thoroughly discussed by Bear (1972). The U.S. Geological Survey model MODFLOW is documented in Harbaugh and McDonald (1996).

3

BUILDING THE
MANAGEMENT
FORMULATION

In this chapter, the steps required to construct a groundwater flow management formulation are presented. Constructing a groundwater management formulation is in many ways similar to constructing a simulation model. First, a formulation statement, which plays the role of the conceptual model, is defined. Next, a specific type of groundwater flow management model is defined that is characterized by the types of objective and constraint functions to be used. Parameters for the management model must be identified, including objective function and constraint coefficient values. Finally, the management model is calibrated and used for design. Constructing the formulation involves a number of subjective decisions. This chapter provides discussion and examples of some of the decisions that need to be made. To simplify the presentation, the formulation is limited to the case of a confined aquifer with steady flow and linear boundary conditions. The methods described here are extended to more complex cases in later chapters.

3.1 FORMULATION STATEMENT

The first step in developing a groundwater flow management formulation is to produce a statement of the intended formulation. The formulation statement answers the questions, what should the design achieve, what are the objectives of the design, and what are the constraints imposed on the design? The formulation may include such objectives as minimize cost, maximize water production, or minimize environmental impact. Constraints might include requirements that costs be maintained within a specified budget, that environmental regulations be satisfied, that stresses be limited by technical considerations, or that total stresses exceed specified production demands. The formulation statement will lead to a complete and unambiguous definition of the objectives and constraints on the particular problem.

The creation of the formulation statement requires interaction between the analyst and decision makers. This interaction is often critical to the success of the project. The process of defining constraints and objectives can lead to a clearer statement of the problem and the elimination of contradictory or redundant objectives. The optimization approach provides a framework in which this interaction can occur. For complex problems, certain considerations such as political, legal, or social issues cannot be directly integrated into the optimization framework. Nevertheless, the process of defining the problem required by the optimization approach can be a significant benefit to any project.

3.2 DECISION VARIABLES, CONSTRAINTS, AND OBJECTIVES

The formulation statement about the problem at hand must be translated into specific mathematical representations. The examples offered in Section 1.3 translate a general conceptual statement into a corresponding mathematical form.

The formulation statement will suggest the characteristics of the problem that require definition as decision variables. In this chapter, the decision variables will be limited to the hydraulic head and the stress at specified locations. When the formulation statement is applied to an aquifer modeled by a numerical scheme, stress locations may be constrained to lie at node points. Discretization of the numerical grid is often organized so that head and stress locations coincide with their corresponding geographic locations.

The breadth of possible objective and constraint functions is enormous and limited only by the imagination of the analyst. Here, we present several general linear forms of objective and constraint functions that can be used to derive a wide variety of specific functions. We also provide examples of the most common constraint and objective forms. These general forms will be used in our subsequent investigations of methods for solving the management formulation. Advanced objective and constraint functions are presented in later chapters.

Constraints may take the form of any function of hydraulic head and stresses. The general form of the constraint functions is indicated in equation (1.2) as a general function of all heads and all stresses. However, practical experience has shown that a wide variety of useful constraints can be derived from a linear combination of heads and a linear combination of stresses. In this chapter, we limit the presentation and analysis to these linear forms.

3.2.1 CONSTRAINTS ON STRESS

In constructing the optimization formulation, n specific locations will be considered for application of stress. A general form of constraints on stress is a bound on a linear combination of stresses at these n locations. This constraint form is written as

$$\alpha^q_{1,k} q_1 + \alpha^q_{2,k} q_2 + \cdots + \alpha^q_{j,k} q_j + \cdots + \alpha^q_{n,k} q_n + \beta^q_k \geq 0 \qquad (3.1)$$

where the $\alpha^q_{j,k}$ and β^q_k are specified coefficients for the kth constraint and the subscripts on the stress variables are location indices in the domain. The variable q_j may represent either extraction or recharge. The general constraint (3.1) is repeated as many times as needed, with different coefficient values, to describe all the constraints on stresses. Following are several examples of specific constraints, and their physical interpretation, that can be derived from the general form (3.1).

Stress Bounds

Constraints are commonly placed upon withdrawal and recharge rates to express bounds on aquifer yield or limitations on the rate at which water can be recharged. Stress bound constraints are represented by retaining a single stress term from (3.1)

$$\alpha^q_{j,k} q_j + \beta^q_k \geq 0 \qquad (3.2)$$

and appropriately defining coefficients. Upper and lower bounds on stress at location j can be expressed as

$$q_j \leq q^u_j \qquad (3.3)$$

$$q_j \geq q^l_j \qquad (3.4)$$

where q^u_j and q^l_j are the maximum and minimum extraction or injection rates that are allowed at location j. These constraints may be repeated for all n stresses. Generally, upper bounds on stress are selected so as to ensure that the surrounding aquifer can accommodate the specified stress. Lower bounds are often dictated by practical considerations and by use of the formulation for location selection as described further in Section 7.2.2.

Bounds on Total Stress

It is often useful to impose bounds on the total stress activity in the aquifer, or in some portion of the aquifer, for purposes of meeting water supply demands or limiting water withdrawals. This can be accomplished by summing the stress and defining the coefficients of (3.1) appropriately. These constraints can take the form

$$q_1 + q_2 + \cdots + q_j + \cdots + q_n \geq Q_k^l \tag{3.5}$$

$$q_1 + q_2 + \cdots + q_j + \cdots + q_n \leq Q_k^u \tag{3.6}$$

where Q_k^l and Q_k^u are, respectively, lower and upper bounds on total stress. If stress at some locations is restricted to recharge only, then constraints of the form of (3.1), with appropriate coefficients, may serve to control the ratio of stress and recharge.

3.2.2 CONSTRAINTS ON HEAD

In constructing the optimization formulation, l specific locations will be considered for observation of head response to stress. A general form of constraint on hydraulic head is a linear combination of these l heads of the form

$$\alpha_{1,k}^h h_1 + \alpha_{2,k}^h h_2 + \cdots + \alpha_{i,k}^h h_i + \cdots + \alpha_{l,k}^h h_l + \beta_k^h \geq 0 \tag{3.7}$$

where the coefficients $\alpha_{i,k}^h$ and β_k^h are specified for the kth constraint and the subscript on the heads indicates a specified observation location. The l heads are those that result from stress at the locations and rates determined by the design process; hence, the heads are implicitly functions of the n stresses. Several examples of specific head constraints are presented in the following, all of which can be derived from the general form (3.7).

Head Bound Constraints

The simplest form of head constraint is a linear function of a single head value. This form of constraint is represented by

$$\alpha_{i,k}^h h_i + \beta_k^h \geq 0 \tag{3.8}$$

By appropriate rearrangement and definition of coefficients, upper and lower bounds on head can be formed as

$$h_i \leq h_i^u \tag{3.9}$$

$$h_i \geq h_i^l \tag{3.10}$$

where h_i^u and h_i^l are, respectively, the upper and lower bound values. Upper bounds on head may be used to control mounding of the piezometric surface. Lower bounds on head may be used to control excessive drawdown as a surrogate for controlling subsidence or ensuring that the piezometric surface remains within a given geologic unit.

Head Difference Constraints

With two head terms retained and appropriate definition of the coefficients, (3.7) yields the head difference constraint

$$h_{k_1} - h_{k_2} \geq h_k^d \tag{3.11}$$

where h_k^d is the specified bound on head difference for the kth constraint and h_{k_1} and h_{k_2} are heads at locations indexed by k_1 and k_2. Constraints of this form can be used to force a gradient in the hydraulic flow field, either horizontally or vertically.

If both sides of (3.11) are divided by the distance, Δx, that separates locations k_1 and k_2, then the head difference constraint can be interpreted as a constraint on hydraulic gradients in a specified direction of the form

$$\frac{h_{k_1} - h_{k_2}}{\Delta x} \geq \frac{h_k^d}{\Delta x} \tag{3.12}$$

Written in this form, it is clear that the head difference constraint simply requires that a component of the gradient vector in the direction defined by locations k_1 and k_2 be greater than the specified value. Finally, (3.12) can be multiplied by the appropriate hydraulic conductivity, K, between points k_1 and k_2 to yield

$$K\left(\frac{h_{k_1} - h_{k_2}}{\Delta x}\right) \geq K\frac{h_k^d}{\Delta x} \tag{3.13}$$

which is now in the form of a velocity or flux requirement. The relationships between head and velocity constraints in multiple dimensions are expanded upon in Section 6.4.

Boundary Flux Constraints

Constraints may be required to control flow between the aquifer and adjacent water bodies. Flux at a single point can often be described by

$$C_i(h_i - H_i) \tag{3.14}$$

where h_i is the head at a point in the aquifer and near the boundary, H_i is the head at a point in the adjacent water body, and C_i is a conductance between the two points. If, for example, F_i is a specified flux that must be maintained, then a constraint of the form

$$C_i(h_i - H_i) \geq F_i \tag{3.15}$$

can be proposed. This constraint differs from (3.13) because it is presumed that the boundary head, H_i, is known. Equation (3.15) can be rewritten as

$$h_i \geq \frac{F_i}{C_i} + H_i \tag{3.16}$$

which highlights the fact that it is of the form of a simple bound on head.

It may be necessary to place constraints on the boundary flux over a region of the aquifer. Such a constraint can take the form of the sum of individual fluxes as

$$\sum_{i=1}^{l} C_i(h_i - H_i) \geq F \tag{3.17}$$

which can be rearranged to the linear form (3.7).

3.2.3 OBJECTIVE FUNCTION

A general linear form of the objective function is

$$f = \beta + \sum_{j=1}^{n} \alpha_j q_j + \sum_{i=1}^{l} \gamma_i h_i \tag{3.18}$$

where α_j, β, and γ_i are specified constants. The constant term, β, is included here to provide the opportunity to enhance the physical meaning of the objective function value. However, when optimizing the objective function, constant terms such as β are irrelevant and can be ignored because changes in the values of the decision variables will not affect these terms.

Optimizing Flow

Perhaps the most common form of objective function is a simple sum of stresses of the form

$$f = \sum_{j=1}^{n} \alpha_j q_j \tag{3.19}$$

If the objective is to maximize water withdrawals, then the coefficients may be set to one and f maximized. If the objective is to minimize operational cost, then f may be minimized with the coefficients representing the cost per unit rate of stress at each location, j.

Optimizing Water Levels

Another common form of objective function is the weighted sum of water levels

$$f = \sum_{i=1}^{l} \gamma_i h_i \qquad (3.20)$$

Maximizing (3.20) for locations i that are coincident with extraction facilities can be used to minimize extraction costs by minimizing the required lift distance. An alternative form for lift cost representation is presented in Section 8.2.1. Minimizing (3.20) when the coefficients are positive will cause maximum drawdown.

A similar objective function can be used for subsidence control problems where the coefficients, γ_i, might represent the unit subsidence per unit head drop or a measure of economic impact as a result of subsidence caused by a unit drop in head. The objective function for this problem takes the form

$$f = \sum_{i=1}^{l} \gamma_i \left(h_i^0 - h_i \right) \qquad (3.21)$$

where h_i^0 is the initial head in the aquifer such that $(h_i^0 - h_i)$ represents the drawdown at location i. Equation (3.21) can be rearranged to appear as the general form in equation (3.18) where

$$f = \beta + \sum_{i=1}^{l} (-\gamma_i) h_i \qquad (3.22)$$

and the constant term $\beta = \sum \gamma_i h_i^0$. Minimizing (3.22) will cause heads to be as high as possible and minimize the total impacts of subsidence.

3.2.4 GENERAL LINEAR MANAGEMENT FORMULATION

Combining the objective and constraints that have been introduced, the general linear formulation can be expressed as

$$\text{minimize} \quad f = \beta + \sum_{j=1}^{n} \alpha_j q_j + \sum_{i=1}^{l} \gamma_i h_i \qquad (3.23)$$

$$\text{such that} \quad \alpha_{1,k}^q q_1 + \alpha_{2,k}^q q_2 + \cdots + \alpha_{j,k}^q q_j + \cdots + \alpha_{n,k}^q q_n + \beta_k^q$$
$$+ \alpha_{1,k}^h h_1 + \alpha_{2,k}^h h_2 + \cdots + \alpha_{i,k}^h h_i + \cdots + \alpha_{l,k}^h h_l + \beta_k^h \geq 0,$$
$$\text{for } k = 1, \ldots, m$$

where m constraints are specified. Constraints on stress and heads can be defined by appropriate selection of nonzero values for the $\alpha_{j,k}^q$ and $\alpha_{i,k}^h$ coefficients.

As noted in Chapter 1, hydraulic head depends on stress, which implies that this linear formulation can be rewritten strictly in terms of stresses.

Such an explicit representation of the formulation dependence on stresses is required for solution and is developed in Chapter 4. In this chapter, the heads are retained to allow direct use of objective and constraint terms that contain heads.

3.2.5 SUCCESSFUL FORMULATION CONSTRUCTION

Successful construction of optimization formulations requires both an understanding of the physical interpretation of the objective and constraints and the ability to anticipate the mathematical impact of the objective and constraints on the form of the solution. Improper design of formulations can lead to problems that have no solution or have solutions that are physically meaningless. Identifying, *a priori*, the presence of these characteristics requires attention to the relationship between objective and constraints and a consideration of how the optimization algorithm will respond to the formulation.

It is important to understand that the optimization formulation contains no inherent logic about the physical reality that it is intended to represent. The objective and constraint functions, along with the relationship between stress and head provided by the simulation model, are the only means by which the optimizer can determine the physical practicality of specific solutions. Lack of consideration of these issues in constructing formulations may produce unintended results. The ability to anticipate the form of the solution is best learned by example. In this section, numerous examples are presented that highlight different aspects of successful formulation construction.

A case in which the behavior of the solution guides the selection of constraints is provided by the subsidence control formulation presented in Section 1.3.3. In this example, a lack of bounds on the stresses could lead to recharge at unintended locations. To avoid such an outcome, the stresses and heads should be bounded.

EXAMPLE OF INCOMPLETE CONSTRAINTS

In the subsidence control example in Section 1.3.3, the objective is to minimize subsidence (through maximizing heads) while meeting water quantity demand, Q. The formulation is stated as

$$\text{maximize} \quad f = \sum_{i=1}^{3} h_i \tag{3.24}$$

$$\text{such that} \quad \sum_{j=1}^{4} q_j \geq Q \tag{3.25}$$

Physical intuition suggests that we would not want to pump more than is necessary to meet the demand because this will produce additional drawdown. Then why is it not necessary to use an equality constraint for meeting water quantity demands? Our physical intuition is reflected in the simulation model, which, when stress is increased, will increase drawdown and reduce the heads. Hence, we can be assured that the solution to this problem will be one in which stress is no more than absolutely needed and that constraint (3.25) will be satisfied as an equality. Further examination of this formulation reveals that it may be incomplete and may yield undesired results. Consider that, in order to maximize heads, the solution may choose to recharge water at some locations (counted as a negative extraction) and extract additional water at other locations to make up the difference required by constraint (3.25). Such a solution might be mathematically possible but undesirable. Avoiding this potential outcome requires adding additional constraints of the form

$$q_j \geq 0, \quad j = 1, \ldots, 4 \tag{3.26}$$

to eliminate recharge or adding constraints of the form

$$h_i \leq h_i^u, \quad i = 1, \ldots, 3 \tag{3.27}$$

to limit explicitly the magnitude of head.

Another example of solution behavior guiding constraint selection can be shown with the saltwater intrusion formulation presented in Section 1.3.4. In this example, recharge would never be selected without the lower head bound constraints. A properly constructed formulation may constrain the decision variables and objective function both directly and indirectly.

EXAMPLE OF COMPLETE CONSTRAINTS

The saltwater intrusion formulation is restated here as

$$\text{maximize} \quad f = \sum_{j=1}^{5} q_j \tag{3.28}$$

such that

$$h_1 \geq h_1^l \tag{3.29}$$

$$h_2 \geq h_2^l \tag{3.30}$$

$$h_3 \geq h_3^l \tag{3.31}$$

$$0 \leq q_j \leq q_j^u, \quad j = 1, \ldots, 3 \tag{3.32}$$

$$q_j \leq 0, \quad j = 4, 5 \tag{3.33}$$

At first glance, maximizing total stress would seem to suggest that stress should go to infinity. However, stress is constrained in several ways. First, there are direct constraints on individual pump rates. If these were the only constraints, then the solution would be to pump all extraction wells at their upper bounds and recharge nothing. However, the presence of the head bounds will presumably prevent this outcome (depending on specific constraint values). Increasing stress tends to decrease head and move head to lower bounds. Hence, we would anticipate that one or more of the head bound constraints would be satisfied as equalities.

As the two preceding examples illustrate, the constraints and objective function drive the optimization algorithm to a specific solution. Anticipation of the likely form of the solution to a management formulation is aided by recognizing that there are only four possible outcomes for a problem of the form (3.23). A linear management formulation can have a single optimal solution, be infeasible, be unbounded, or have multiple optimal solutions. Examples of formulations that do not result in a single optimal solution are presented in the following and suggestions are provided for means of identifying problems in the formulation.

Infeasible Problems

A formulation that contains constraints that cannot be simultaneously satisfied by any one set of stresses is said to be infeasible. Careless formulation of the groundwater flow management problem can produce inadvertent infeasible problems. Infeasibility can also result from well-posed formulations in which the constraints are incompatible with the physical limitations of the system.

EXAMPLE OF AN INFEASIBLE PROBLEM

Consider the construction dewatering and wetland protection problem depicted in plan view in Figure 3.1. Suppose the problem is to lower the water table within the construction zone while maintaining head values in the wetland area. One approach is to place candidate extraction and injection wells around the construction area as shown in Figure 3.1. If the extraction and injection wells are placed too close together, the problem may be infeasible. This would result if the head bound constraints in the wetland area could not be met because the downgradient extraction wells capture most of the injected water. A more realistic well placement would have the injection wells adjacent to the wetland so that the downgradient extraction wells do not capture a large portion of the injected water but instead withdraw water from the construction area.

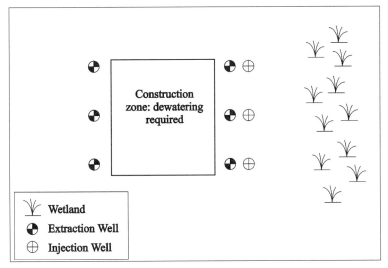

FIGURE 3.1 Placing injection wells adjacent to extraction wells may lead to an infeasible problem. Head in the wetland cannot be maintained if extraction wells capture most of the injected water.

Infeasible problems may also arise in properly formulated problems because of competing or incompatible design criteria. For example, Figure 3.2 depicts a cross section of another construction dewatering problem with a nearby wetland. Assume that regulatory requirements prohibit injection near the wetland so that only extraction wells can be considered as candidate wells. If the aquifer is sufficiently transmissive, then it may be physically impossible to lower water levels at the construction site without affecting water levels in the wetland. This problem is well posed from an optimization viewpoint. However, the design

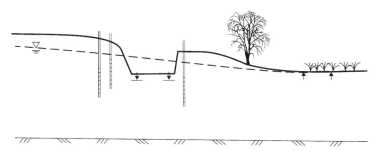

FIGURE 3.2 The hydraulic properties of the aquifer may be such that it is physically impossible to meet the head bound constraints for both construction dewatering and wetland protection.

criteria that are imposed are incompatible with the physical character istics of the aquifer.

Identifying that no feasible solution exists may be of significant value to the analyst and decision maker. Formulations that have no feasible solutions conclusively prove the incompatibility of the design criteria when the constraints that are posed accurately reflect these criteria.

Unbounded Problem

An unbounded problem is one in which the optimal objective function value is either positive infinity or negative infinity. Such a solution will usually have no physical significance but will instead indicate that the formulation is poorly posed.

EXAMPLE OF AN UNBOUNDED PROBLEM

Consider the formulation that seeks to maximize extraction subject to upper bound head constraints:

$$\text{maximize} \quad f = \sum_{j=1}^{n} q_j \tag{3.34}$$

$$\text{such that} \quad h_i \leq h_i^u \qquad i = 1, \ldots, l \tag{3.35}$$

The solution to this problem is positive infinity because there are no upper bounds on the stresses. Depending on the numerical simulation model used and the way in which it is linked to the optimizer, the simulator may fail before the optimizer reaches infinity.

To avoid unbounded problems, ensure that every variable is bounded. In addition to upper and lower bounds on variables, bounds may be imposed indirectly through the physical constraints represented by other variables. For example, a lower bound on head will indirectly provide an upper bound on extraction because more stress produces more drawdown.

Multiple Optima

In rare circumstances, the general linear formulation (3.23) may have multiple optimal solutions. The values of the decision variables in these solutions will be different but the objective function values will be identical. Hence, all of the optimal solutions are equally good. Multiple optima arise when a constraint that is satisfied as an equality at the solution is collinear with the objective function.

EXAMPLE OF MULTIPLE OPTIMA

Consider a water supply problem in which the maximum stress from two wells is sought subject to constraints on water demand, treatment plant capacity, and drawdown in the aquifer. The problem is stated mathematically as

$$\text{maximize} \quad f = q_1 + q_2 \tag{3.36}$$

such that

$$q_1 + q_2 \leq Q^u \tag{3.37}$$

$$q_1 + q_2 \geq Q^l \tag{3.38}$$

$$h_1 \geq h_1^l \tag{3.39}$$

$$h_2 \geq h_2^l \tag{3.40}$$

where Q^u is the treatment facility capacity, Q^l is the minimum water demand, and h_1 and h_2 are the heads at the two extraction wells. Notice that the objective function and first two constraints describe parallel lines. When the objective function equals Q^u, then the objective function and first constraint are collinear. Assuming that the plant capacity constraint limits the solution, then the multiple optimal solutions are the set of feasible stresses that satisfy the first constraint as an equality.

Any of the multiple optimal solutions can be adopted as the optimal solution. From a practical viewpoint, the selection of the best among the multiple solutions can either be arbitrary or based on considerations not incorporated in the formulation.

3.3 PARAMETERS

The parameters of the groundwater flow management formulation are the number of candidate stress locations, n, the number of head observation locations, l, the number of constraints, m, and all of the coefficients, β, α_j, γ_i, β_k^q, $\alpha_{j,k}^q$, β_k^h, and $\alpha_{i,k}^h$. Specification of all parameters is required to define the problem fully. As noted in the previous section, the type of objective or constraint to be employed is determined by proper selection of coefficient values for the objective function and constraints. However, additional considerations come into play when selecting the specific values of these coefficients and the values of the parameters l, m, and n.

Rates of stress, q_j, are the primary decision variables. The parameters required to describe q_j are the number of locations to be considered for stress, n, and the specific locations in the aquifer that will be considered.

Hydraulic head, h_i, is the primary dependent variable. The parameters required to describe h_i are the number of observation locations, l, and the specific locations for observation in the aquifer. Specification of each constraint requires two elements: specifying the stress or head locations that will be constrained and selecting the coefficient values associated with those constraints. In this section, issues that arise in parameter selection are discussed and examples are provided to illustrate the concepts.

3.3.1 STRESS DECISION VARIABLES

The optimization formulation is constructed with n locations at which stress can be applied to achieve the system state defined by the constraint functions. These n locations are referred to as candidate locations to emphasize the fact that the solution of the optimization formulation can draw from these locations for active stress points. It is often the case, when sufficient candidate locations are provided, that the solution of a given formulation includes the selection of zero stress at some of the candidate locations.

The tendency of a solution to contain zero-valued stress at certain candidate locations is demonstrated in this simple example. Consider a confined aquifer with constant transmissivity, T, a constant head boundary at a distance R, and n wells, each with stress rate q_j. Using the Theim equation and the principle of superposition and assuming that the same value of R can be applied to each well, the drawdown at a single point in response to stress at multiple wells can be computed as

$$s(r) = \sum_{j=1}^{n} \frac{q_j}{2\pi T} \ln\left(\frac{R}{r_j}\right) \tag{3.41}$$

where s is the drawdown and r_j is the distance between well j and the location at which drawdown is being predicted.

Consider that this model is applied to a problem in which drawdown at a single point is constrained to be greater than a specified value (a dewatering problem) and total stress is to be minimized. Such a problem setup will ensure that the single drawdown constraint will be binding. Hence, the resulting optimization problem can be written as

$$\text{minimize} \quad f = \sum_{j=1}^{n} q_j \tag{3.42}$$

$$\text{such that} \quad \sum_{j=1}^{n} \left[\frac{\ln\left(\frac{R}{r_j}\right)}{2\pi T}\right] q_j = s^* \tag{3.43}$$

The coefficients on the stresses depend only on the distance from the pumping well to the constraint. The single well closest to the constraint will have the largest coefficient. Hence, the solution is to pump the closest well at a rate that will yield the desired drawdown. If upper bounds are present on stresses, then the solution would be to pump, to their capacity, each of the wells closest to the constraint point. Although multiple constraints and heterogeneities will complicate the solution, the conclusion remains that the solution will consist of the subset of wells that most directly satisfy the constraints and that some of the stresses will be zero.

It is possible to associate the selection of nonzero stress with the selection of the optimal location of a new stress point. Hence, the optimization formulation can be viewed as simultaneously accomplishing two tasks: identifying optimal stress locations from a set of candidate locations and identifying the optimal magnitude of stress at those locations.

It should be emphasized that solving the optimal location problem in this fashion is approximate only. The candidate locations are fixed points on the numerical grid. The true optimal locations may be at points not associated with grid nodes and inadequate candidate locations may be provided to identify the true optimal locations. The difference between the approximate and true optimal location problem solution depends, in part, on the density of candidate locations. As the number of candidate locations is increased, the opportunity for solving the true location problem also increases. Hence, numerous points (e.g., every node in a numerical grid) might be specified as candidate locations.

Several problems arise if an excessive number of candidate locations are used. The first problem with too many candidate locations is the computational burden. As discussed in more detail in Section 4.5, the computational effort to solve a groundwater management problem is directly related to the number of candidate locations used. The second problem with excessive candidate locations concerns solution stability under perturbation. When a formulation is used to select stress locations, a desirable property is that the set of optimal locations does not dramatically change when minor changes are made to the constraints. This property may not be present when multiple, densely spaced candidate wells are available.

EXAMPLE OF STRESS LOCATION STABILITY

To illustrate the concept of solution stability, consider the example of a confined aquifer with a transmissivity of 140 m^2/day and an initial

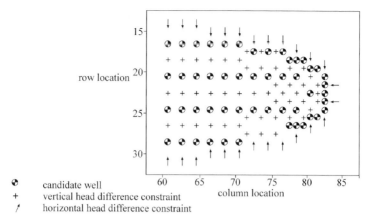

FIGURE 3.3   Problem setup showing candidate well locations and vertical and horizontal head difference constraint locations.

uniform gradient of 0.0207. The simulation model uses a grid with a uniform spacing of 10 meters. Head difference constraints are imposed in both the horizontal and vertical directions to form a capture zone for the containment of contaminated groundwater. Figure 3.3 depicts the location of the head difference constraints and candidate wells in the portion of the domain indicated by the row and column location indices. Solution of the problem yields the set of active stress points depicted in Figure 3.4. A 50% reduction of the head difference constraint value at the location indicated in the upper left corner of Figure 3.4

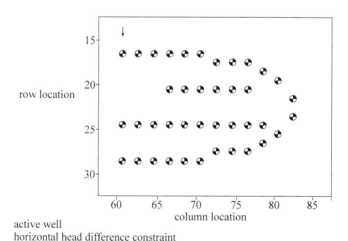

FIGURE 3.4   Optimal well locations from problem depicted in Figure 3.3. Perturbation of the constraint shown yields the changes in well locations shown in Figure 3.5.

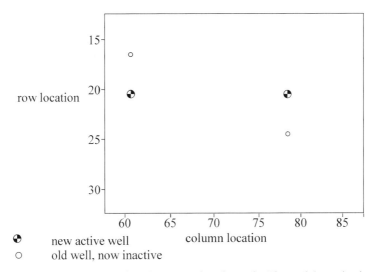

● new active well
○ old well, now inactive

FIGURE 3.5 Perturbing the constraint shown in Figure 3.4 results in changes in four well locations.

yields the changes in active wells shown in Figure 3.5. Two active wells have been eliminated and two new locations have been activated. Note that two of these changes occur some distance from the location of the change in constraint. This solution would be considered unstable. Stability can be improved for this example by decreasing the number of

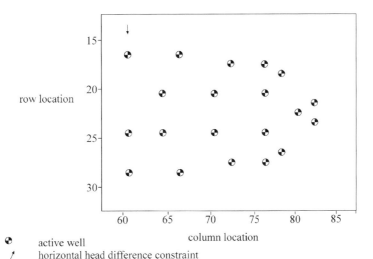

● active well
↗ horizontal head difference constraint

FIGURE 3.6 Optimal well locations when 60% of the candidate wells shown in Figure 3.3 are eliminated from consideration. Perturbation of the constraint shown does not result in any change in the optimal well locations.

candidate wells. Eliminating 60% of the candidate wells and re-solving the problem with the original parameter values yields the solution depicted in Figure 3.6. A 50% reduction of the same head difference constraint indicated yields no changes in the set of active wells.

General insights into the source of the behavior depicted in this example can be developed by studying the solution algorithm and the sensitivity and range analysis discussed in Chapters 4 and 5. However, specific identification of instability is hindered by its dependence on the constraints imposed, the set of candidate locations used, and site-specific hydrogeologic conditions. The preceding example suggests that it is unwise to use a high density of candidate locations. However, use of only a few candidate locations requires *a priori* selection of location and eliminates the value of the optimization algorithm to solve the location selection problem. One heuristic approach to addressing this problem is to provide numerous candidate locations to solve the initial location problem. Then, in subsequent perturbation runs, limit the number of candidate locations to those selected in the initial solution.

The use of multiple candidate locations as a means of selecting optimal locations for stress also has implications for setting lower bounds on stress. When using multiple candidate locations, the lower bound for extraction stresses and the upper bound for recharge stresses should be set to zero. If these bounds are set to nonzero values, then all candidate locations are forced to operate at a nominal level, eliminating the opportunity for the formulation to set selectively nonzero stress. The disadvantage of using a zero bound is that the solution may contain some locations that are assigned rates that are unrealistically small. Methods for imposing a requirement that the stress be either zero or bounded between two nonzero values require the use of integer variables and are discussed in Chapter 7.

3.3.2 CONSTRAINT LOCATIONS AND VALUES

Specifying constraint location involves identifying the specific location or locations where the constraint will be applied. The constraints are often surrogates for the actual design criteria to be satisfied. The criteria are often defined in terms of hydraulic control over an area or volume of the aquifer. If the area to be controlled is sufficiently large, then multiple constraints will be required to satisfy the design criteria. Hence, the question of the number of constraints to use (or the density of constraint placement) arises.

A low constraint density may mean that the original design criteria are not satisfied. High constraint density comes at increased computational cost and may produce instability in the optimal solution similar to that identified for candidate stress locations. In summary, constraint placement is governed by three considerations: 1) satisfying the physical requirement, 2) limiting computational effort, and 3) enhancing solution stability. The first consideration generally implies the use of a higher density of constraints, while the second and third considerations imply the use of fewer constraints.

EXAMPLE OF CONSTRAINT DENSITY

To gain insight into the impact of constraint density and the use of surrogate constraints to represent design criteria, consider a dewatering example similar to that posed in Section 1.3.2. The area enclosed by the thick black line in Figure 3.7 must be dewatered for construction purposes. The aquifer is modeled as confined with a transmissivity of 50 ft^2/day and an initial uniform gradient of 0.007 feet per foot with flow from left to right. The model domain is discretized into 600 finite difference grid cells, each measuring 100 feet by 100 feet. The specific design criteria are that the water levels within the L-shaped excavation area must be maintained beneath a specified level, in this case 50 ft. Wells may be placed around the perimeter of the excavation. Surrogate design criteria for this case are that the head at each node that lies within the portion of the grid containing the simulated excavation should have an upper bound constraint imposed. Assuming that the node placement is sufficiently dense, a scheme that satisfies these constraints will also satisfy the physical objective.

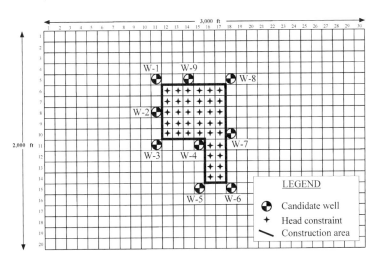

FIGURE 3.7   Construction dewatering example; the head at each constraint location must be maintained below 50 ft. The finite difference grid is shown for reference.

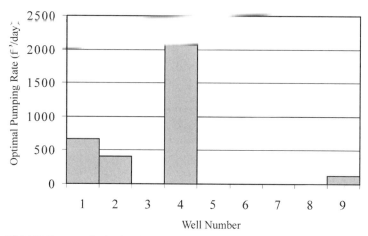

FIGURE 3.8   Optimal solution for construction dewatering example.

The optimal solution to this problem is shown in Figure 3.8; the piezometric surface resulting from this pumping scheme is depicted in Figure 3.9 and confirms that the constraints have been satisfied at the center of all cells within the enclosed area. Although the solution obtained is adequate, we might ask whether this density of constraints is actually needed. From basic drawdown theory we can predict that if the wells extract water at a rate sufficient to satisfy head upper bound constraints in the center and around the perimeter of the excavation, then all intervening constraints will also be satisfied. Basing our constraint selection on these theoretical considerations, we can reduce the num-

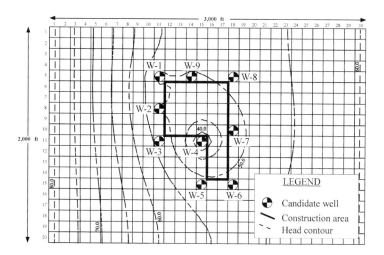

FIGURE 3.9   Piezometric surface upon application of optimal pumping rates. All simulated heads within the construction area are at or below 50 ft.

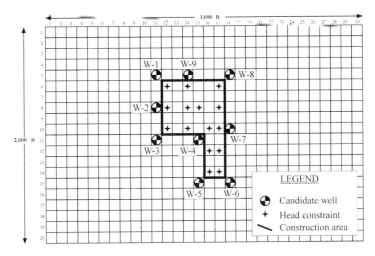

FIGURE 3.10 Reduced constraint set that results in the same optimal solution.

ber of constraints to include only node locations around the perimeter and in the center of the excavation. In fact, the reduced constraint set depicted in Figure 3.10 is sufficient to obtain the same optimal solution as determined before, with less computational effort.

Another consideration in constraint specification is the selection of the specific values used for constraints. These values depend on the type of constraint, the way in which the constraint is used, and the placement of the constraint. In many applications there is a trade-off between density of constraint placement and the constraint values used. As the constraint density increases, the constraint values can be set to represent the desired constraint directly. At coarser constraint densities, both the constraint locations and values are considered surrogates that indirectly accomplish the design criteria.

EXAMPLE OF CONSTRAINT MAGNITUDE

As an illustration of the relationship between constraint value, stress, and constraint location, consider an example involving head difference constraints used to cause a reversal in flow. A well is placed in a regional flow field and a horizontal gradient constraint is placed some distance downgradient. In Figure 3.11a the constraint is set to a relatively large value, as indicated by the slope of the arrow. Substantial stress is required to satisfy the constraint and the capture zone is extended some

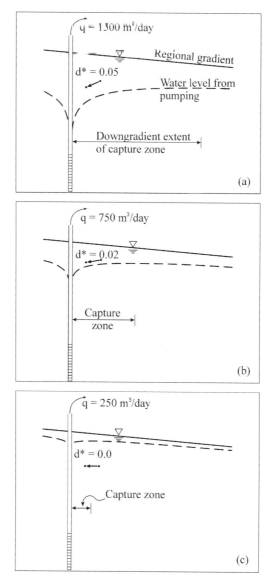

FIGURE 3.11   The value used to constrain the head difference affects the optimal pumping rate and the size of the capture zone.

distance downgradient from the constraint location. In Figure 3.11b, a smaller gradient requirement yields smaller drawdown and less propagation of the capture zone. In general, the larger the constraint value the more stress will be required to meet the constraint. When candidate wells are placed far from the constraints, large constraint values

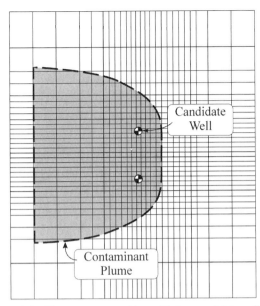

FIGURE 3.12 Plume capture problem with two candidate wells. Head difference constraints are used as surrogates to represent an implied plume capture goal.

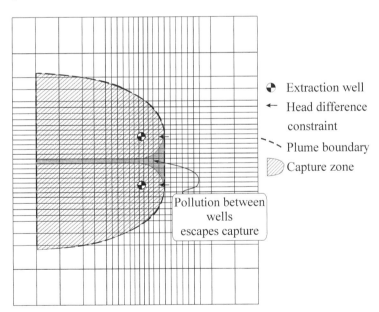

FIGURE 3.13 Two zero-gradient constraints are not sufficient to capture the entire plume.

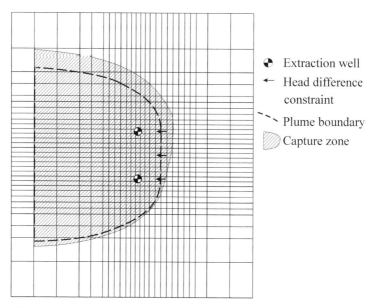

FIGURE 3.14   Three zero-gradient constraints result in a capture zone that envelops the entire plume.

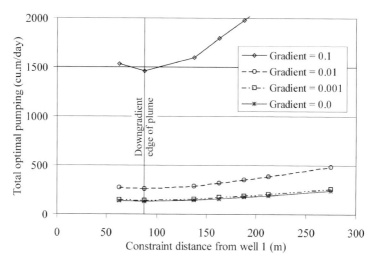

FIGURE 3.15   Total optimal pumping for different head difference constraint locations and values. All formulations include three head difference constraints.

can yield infeasible solutions. Finally, in Figure 3.11c, the gradient is specified to be zero at the constraint location. In this case, the capture zone ends at the constraint location and the least stress is required.

The approach of using a zero gradient requirement to control the location of a capture zone boundary may not be successful for multiple wells or constraints because there is no guarantee that capture will be achieved in the locations adjacent to or between these constraints. This idea is demonstrated with the two-dimensional plume capture example shown in Figure 3.12. Here, two candidate wells are placed in the downgradient region of a contaminant plume. Head difference constraints are placed downgradient of the plume in order to effect flow reversal to the extraction wells. The question arises as to the number of constraints needed and their appropriate values. Figure 3.13 depicts the capture zone of the optimal solution when only two constraints are used and their values are not sufficient to capture the plume. When a third constraint is added, the plume is indeed captured, but the capture zone is larger than the plume, as shown in Figure 3.14. In fact, the stress required to satisfy the constraints changes substantially depending on both constraint value and constraint location, as shown in Figure 3.15.

The preceding examples offer some insight into the nature of the trade-offs inherent in using constraints as surrogate design criteria. The exact nature of these trade-offs is problem dependent and should be investigated during model calibration, which is discussed in Section 3.4.

3.3.3 RELATIONSHIP BETWEEN STRESS AND CONSTRAINT LOCATIONS

The placement of candidate stresses relative to constraints is also an important issue. A common implementation of constraint placement and candidate stress placement is to impose many constraints along with many candidate stresses. The expectation is that only several candidate locations will be selected to satisfy the constraints. However, this intention can be thwarted by placing candidate locations too close to constraints. This can activate locations with unrealistically low stresses that satisfy the nearby constraint but are not part of a larger solution.

EXAMPLE OF STRESS AND CONSTRAINT PLACEMENT

As an illustration of the relationship between stress and constraint locations consider the four-well placement problem shown in

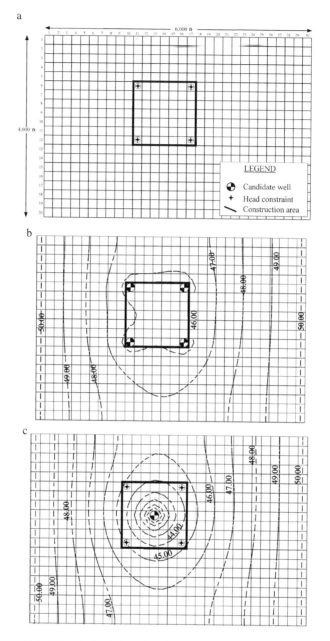

FIGURE 3.16   (a) If only four drawdown constraints are used as surrogates for dewatering the entire square region, what is the most effective well placement? (b) One well at each constraint location does not dewater the entire square region. (c) One well in the middle of the square can produce the desired drawdown.

Figure 3.16a The piezometric surface is initially flat at 50 feet and the region within the square must be dewatered to 45 feet. If four head constraints are placed at the corners of the square to act as surrogates for required drawdowns, then a single well in the middle will do better at dewatering the entire square region than one well at each constraint. The potentiometric surfaces resulting from the optimal pumping solutions for these two cases are shown in Figures 3.16b and c. Note that the result depicted in Figure 3.16b is a trivial solution. Each well pumps just enough to satisfy the head constraint at the same location. Heads in the remainder of the square do not fulfill the original design goal of dewatering the entire region.

When constructing a management formulation, the modeler must carefully consider the behavior of the optimizer while also providing hydrogeologic insight to anticipate the possible interactions between stress and constraint location.

3.3.4 OBJECTIVE FUNCTION PARAMETERS

The objective function can be used to represent such varied quantities as the cost of system operation, the value of resource utilization, or the impact of applied stress on the aquifer. The parameters of the objective function along with the functional form of the objective function translate the value of the decision variables into the quantity of interest. The definition of objective function parameters can be simplified by recognizing several principles of function minimization.

The solution to the problem minimize $[f(\mathbf{q})]$ will be identical to the solution of the problem minimize $[f(\mathbf{q}) + \varphi]$, where φ is a specified constant. The implication of the preceding statement is that any terms in the objective function that do not depend on the decision variables can be eliminated from the formulation for the purpose of finding the optimal solution. These terms may be returned to the objective function to compute a meaningful numerical value of the objective once a solution is identified.

Similarly, the solution to the problem minimize $[f(\mathbf{q})]$ will be identical to the solution of the problem minimize $[\varphi f(\mathbf{q})]$, where φ is a specified constant. For example, the common objective to minimize stress costs takes the form

$$\text{minimize} \quad f = \sum_{j=1}^{n} c_j q_j \tag{3.44}$$

where the objective function coefficients, c_j, would typically have dimensions of monetary units/stress rate, so that the objective function has monetary units. However, if the coefficients all have the same value then this objective will produce an identical stress solution to the objective

$$\text{minimize} \quad f = \sum_{j=1}^{n} q_j \tag{3.45}$$

In addition, it can be observed that, even when coefficients differ, they can all be multiplied by a constant and the optimal solution will not change. Thus, the absolute unit costs of stress need not be determined; instead only the relative costs are required.

3.4 MANAGEMENT MODEL CALIBRATION

Calibration is one of the fundamental tools in building models and is defined as the process of modifying model characteristics and parameters to produce a best match between desired outcomes and model predictions. Calibration is an important step in constructing groundwater simulation models where model predictions are compared with field data. In the context of groundwater flow management models, calibration can be applied to the selection of constraint and objective functions and to the choice of management model parameters. The purpose of calibration is to select management model parameters that produce results that best match the intended design criteria. This is important for parameters that serve to represent the design criteria indirectly. For example, in the case of creating a capture zone, head difference constraints are used as surrogates for plume capture. Here, calibration requires adjusting parameters so that plume capture is obtained. Confirming plume capture might be accomplished with flow path or particle tracking analysis or some other form of postoptimization solution checking.

The construction of the groundwater flow management formulation may be viewed as an iterative process as depicted in the flowchart in Figure 3.17. Here, the formulation statement is defined, constraint and objective functions are selected, parameters are selected, and the problem is solved. It is then confirmed that the solution does indeed satisfy the design criteria. This step is often performed by postoptimal simulation of the aquifer to test further the system response to the solution generated. If the design criteria are not satisfied, it is then necessary to modify the formulation. This may involve adding or subtracting candidate stress points, adding or subtracting constraints, or modifying constraint values.

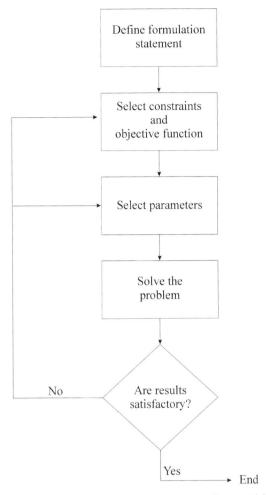

FIGURE 3.17 The process of constructing a hydraulic control formulation.

3.5 MANAGEMENT MODEL VERIFICATION

In modeling analysis, the verification step is intended to confirm that the solution to the mathematical model is, in fact, the correct solution to the original governing equations. This step is particularly important for mathematical models that are solved with numerical methods because approximations are implicit in the solution process. The verification process is generally independent of the particular parameter values used.

The verification concept can also be applied to the management model. It is assumed that the simulation model, upon which the management model depends, has been independently verified to solve the governing equations correctly. Similarly, it is assumed that the optimization solution algorithm has been independently verified to find the optimal solution to a general set of formulations. It remains to verify that the constraint and objective functions are correctly assembled.

Verification of the groundwater flow management model implies confirming that the solution obtained from the management model is both feasible and optimal. This testing can be conducted by simulating the aquifer response to the optimal set of stresses. Simulation model output can be used to confirm that the constraints are satisfied. Optimality can be confirmed by repeated simulation with different stresses. Simulations are conducted by increasing and decreasing each stress in turn. Optimality is confirmed if it can be shown that each stress is either suboptimal (i.e., degrades the objective function value) or produces an infeasibility with respect to the constraints. Many optimization algorithms implicitly perform these verification steps, making manual verification unneccessary.

3.6 MANAGEMENT MODEL VALIDATION

In simulation modeling analysis, the validation step typically involves using the model to simulate the system response to a set of inputs that are different from those that have been used for model construction and calibration. These simulated outputs are compared against measured results to test the ability of the model to simulate system response without any additional parameter modification. The validation step is often used to provide confidence that the model so tested can, in fact, simulate the system under study. Validating a groundwater flow management model is complicated because such models contain three components: the simulation model, the optimization algorithm, and the optimization formulation.

Use of the simulation model by the groundwater flow management model is predicated on the assumption that the simulation model is a reliable predictor of system response. The construction, calibration, and validation (if any) of the simulation model are outside the scope of the management model and are necessary but not sufficient conditions for validation of the management model.

Validation of the optimization algorithm is also outside the scope of the management model. As discussed in Chapter 4, algorithms for solving the formulations described in this book are well developed. In most cases, mathematical proofs verify that the algorithms will converge to the optimal solution (if it exists) for any properly posed problem. Independent validation of the optimization algorithm in the context of management models is not necessary.

The final component of the groundwater flow management model is the particular set of constraint and objective functions that make up the formulation for a specific problem. Validation of the management formulation can take many forms depending on the nature of the formulation. In general, validation implies confirming that a desirable feature of the formulation is retained when one component of the formulation remains fixed and another is modified.

For example, a limited number of constraints may be imposed with the expectation that satisfying some constraints implies satisfying others that are not explicitly stated. In this context, validation would require checking the unstated constraints and confirming that they are indeed satisfied when constraint locations and constraint parameter values remain fixed while inputs to the groundwater flow management model such as candidate stresses or bounds on stress are modified.

It is important to note that it is not possible to provide absolute proof that a strategy produced by use of the management formulation will actually be optimal when applied in the field. One can say that if 1) the simulation model is an exact predictor of system response, 2) the management formulation perfectly represents the design criteria, and 3) the optimization algorithm can find the optimal solution, then the strategy is indeed optimal. The last condition is easily met and the second condition can often be confirmed by appropriate calibration and validation measures. However, practical experience suggests that the first condition can rarely, if ever, be satisfied. The analyst can make the conditional statement that the strategy produced is optimal with respect to the simulation model used and the formulation posed.

3.7 THE WATER SUPPLY EXAMPLE

A simple example is used to illustrate elements of formulation construction. This example will be used to demonstrate solution procedures and methods for result analysis in later chapters. The example consists of an aquifer that discharges to a river under steady flow conditions. Two water supply wells are to be operated in such a way as to maximize the total rate of water withdrawal under steady conditions. Furthermore, the extraction scheme must minimize the amount of water that is drawn from the river as a result of pumping. Well locations relative to the river are depicted in Figure 3.18a. The physical system precludes simulation with an analytical model, so the analysis is to be performed on a numerical grid.

For this water supply example, a formulation statement that seeks simply to maximize withdrawals is incomplete because there is no considera-

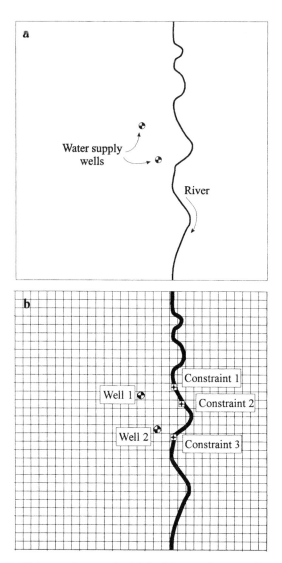

FIGURE 3.18   Water supply example. (a) In this example, we seek to maximize water supply from the two wells while minimizing the amount of water that flows from the river into the aquifer as a result of pumping. (b) The numerical discretization scheme and constraint locations.

tion of river withdrawal concerns. Similarly, simply minimizing river water withdrawal, with no requirements on stress rates, would result in a solution with little or no pumping. The formulation statement of maximizing withdrawals while simultaneously minimizing river water withdrawal is ambiguous. Such a statement would yield a solution that would depend on the relative weightings used on the pumping rates and heads (see Section 5.1.3).

Based on the problem description, two formulation statements are possible. One formulation would be to maximize the water withdrawn from the wells while requiring no withdrawal from the river. The flux between surface water and groundwater is often modeled as a function of the head difference between the two water sources. Therefore, we can restrict withdrawal from the river by constraining the hydraulic head in the aquifer to be at least as high as the river stage. We could place head constraints at all grid cells that contain the river, but a more efficient formulation includes constraints only at critical locations. Because the unperturbed groundwater flow is toward the river, we can expect the downgradient river cell closest to each extraction well to limit the solution. Furthermore, we might expect a river cell between the two extraction wells to provide the critical drawdown by the principle of superposition. The location of the second constraint, shown in Figure 3.18b, was chosen because it is closer to the candidate wells than nearby river nodes yet it is not adjacent to the other constraints. Based on this analysis, the constraints are defined by imposing a requirement on drawdown at three locations on the numerical grid along the river as shown in Figure 3.18b. The constraints require that head in the aquifer not decline below that found in the river. The resulting formulation is

$$\text{maximize} \quad f = q_1 + q_2 \tag{3.46}$$

$$\text{such that} \quad h_i \geq h_i^l \qquad i = 1, 2, 3 \tag{3.47}$$

The three constraints are used as surrogates for the entire river length and their locations are chosen to represent the grid nodes expected to exhibit the largest drawdown as a result of pumping. In this case, the river nodes closest to each of the wells and one in between the wells are expected to exhibit the most drawdown.

An alternative formulation would be to maximize the head near the river (as a surrogate for minimizing withdrawal from the river) while requiring that a minimum amount of pumping be obtained. This formulation takes the form

$$\text{maximize} \quad f = \sum_{i=1}^{3} h_i \tag{3.48}$$

$$\text{such that} \quad q_1 + q_2 \geq Q_{\text{demand}} \tag{3.49}$$

In this formulation, the value of the water demand will determine whether any withdrawals from the river are required.

3.8 NOTES AND REFERENCES

The basic forms of the constraint and objective functions presented in Section 3.2 developed as specific application problems arose. The objective of maximizing total pumping was used in a water supply example (Deninger, 1970), and minimizing total pumping was used for aquifer dewatering (Aguado et al., 1974). The objective of maximizing hydraulic head under a requirement on total pumping was used by Aguado and Remson (1974). Lower bounds on head to limit drawdown were used for a water supply system (Deninger, 1970) and in an agricultural planning application (Maddock, 1973). Upper bounds on head were used in a dewatering example (Aguado et al., 1974). Head difference constraints were used to control the migration of contaminated water in an aquifer (Molz and Bell, 1977).

Although the early development of groundwater flow management models was driven by specific applications, these works are limited by application to small hypothetical problems. With the development of more sophisticated software and higher capacity computers, the application of these methods to field problems has become practical. As in the development of simulation models, the skills needed to construct management formulations effectively are acquired, in part, by studying the efforts of others. The interested reader may pursue some of the references listed for additional insights into the steps involved in constructing formulations. Ahlfeld and Heidari (1994) focus their review paper on applications of these methods. Some examples of applications are plume control problems (Tiedeman and Gorelick, 1993; Ahlfeld et al., 1995), water supply–related applications (Lall and Lin, 1991; Gharbi and Peralta, 1994; Peralta et al., 1995; Barlow et al., 1995, Nishikawa, 1998), and applications to control groundwater levels (Danskin and Freckleton, 1992) and to control saltwater intrusion (Finney et al., 1992; Hallaji and Yazicigil, 1996; Emch and Yeh, 1998).

The stress location stability example and associated figures are drawn from Verdon (1995).

4

SOLVING THE
MANAGEMENT
FORMULATION

In the previous chapter, the general linear form of the optimization problem is stated as

$$\text{minimize} \quad f = \beta + \sum_{j=1}^{n} \alpha_j q_j + \sum_{i=1}^{l} \gamma_i h_i \tag{4.1}$$

$$\text{such that} \quad \alpha_{1,k}^q q_1 + \alpha_{2,k}^q q_2 + \cdots + \alpha_{j,k}^q q_j + \cdots + \alpha_{n,k}^q q_n + \beta_k^q +$$

$$\alpha_{1,k}^h h_1 + \alpha_{2,k}^h h_2 + \cdots + \alpha_{i,k}^h h_i + \cdots + \alpha_{l,k}^h h_l + \beta_k^h \geq 0,$$

$$\text{for } k = 1, \ldots, m$$

In this chapter, the simplex method, a highly efficient, reliable, and widely used method for solving linear optimization problems, will be presented. Use of this algorithm requires that the formulation be stated in the form of a linear program. A linear program is an optimization formulation in which the objective function and all constraints are linear functions of the decision variables. The groundwater flow management problem (4.1) can be transformed to the form

$$\text{minimize} \quad f = \sum_{j=1}^{n} c_j q_j \tag{4.2}$$

$$\text{such that} \quad \sum_{j=1}^{n} a_{kj} q_j \geq b_k, \quad k = 1, \ldots, m$$

The transformation requires explicitly defining heads as a function of stresses.

4.1 MANAGEMENT PROBLEM AS
A LINEAR PROGRAM

In this section we show how many constraints and objectives can be transformed into a linear program of the form (4.2). The primary functional relationship in the formulations described to this point is that between hydraulic head and stress. The response of head at location i to a change in pumping can be described using the Taylor series as

$$h_i(\mathbf{q}) = h_i^0(\mathbf{q}_0) + \sum_{j=1}^{n} \frac{\partial h_i}{\partial q_j}(\mathbf{q}_0)\left(q_j - q_j^0\right)$$ (4.3)

$$+ \sum_{j=1}^{n} \sum_{k=1}^{n} \left[\frac{\partial^2 h_i}{\partial q_j \partial q_k}(\mathbf{q}_0) \frac{\left(q_j - q_j^0\right)\left(q_k - q_k^0\right)}{2!} \right] + \cdots$$

where $\mathbf{q}$ is the vector of new stresses with elements q_j and $\mathbf{q}_0$ is the vector of original stresses with elements q_j^0.

If it is assumed that head is a linear function of stress, as is the case for confined aquifers (see Section 2.3.4), then the second and higher derivatives of h with respect to q are zero and the first derivative of h with respect to q is a constant. This further implies that the first derivative of head is independent of the value of stress at which it is evaluated. It will be most convenient to select all elements of $\mathbf{q}_0$ to be zero. Using these simplifications, the Taylor series reduces to

$$h_i(\mathbf{q}) = h_i^0 + \sum_{j=1}^{n} \frac{\partial h_i}{\partial q_j} q_j$$ (4.4)

The general form of head constraints is the linear combination of heads, repeated here as

$$\alpha_1^h h_1 + \alpha_2^h h_2 + \cdots + \alpha_i^h h_i + \cdots + \alpha_l^h h_l + \beta^h \geq 0$$ (4.5)

Substituting (4.4) for each head yields

$$\alpha_1^h\left(h_1^0 + \sum_{j=1}^{n} \frac{\partial h_1}{\partial q_j} q_j\right) + \alpha_2^h\left(h_2^0 + \sum_{j=1}^{n} \frac{\partial h_2}{\partial q_j} q_j\right) + \cdots$$ (4.6)

$$+ \alpha_i^h\left(h_i^0 + \sum_{j=1}^{n} \frac{\partial h_i}{\partial q_j} q_j\right) + \cdots$$

$$+ \alpha_l^h\left(h_l^0 + \sum_{j=1}^{n} \frac{\partial h_l}{\partial q_j} q_j\right) + \beta^h \geq 0$$

or

$$\sum_{i=1}^{l} \alpha_i^h \left(h_i^0 + \sum_{j=1}^{n} \frac{\partial h_i}{\partial q_j} q_j \right) + \beta^h \geq 0 \tag{4.7}$$

or

$$\sum_{j=1}^{n} \left(\sum_{i=1}^{l} \alpha_i^h \frac{\partial h_i}{\partial q_j} \right) q_j + \sum_{i=1}^{l} \alpha_i^h h_i^0 + \beta^h \geq 0 \tag{4.8}$$

The constraint in terms of heads, equation (4.5), has been converted into a constraint in terms of stresses, (4.8). Use of (4.8) in place of the portion of the general constraints in (4.1) that involve heads yields a formulation entirely in terms of the stress decision variables of the form of (4.2).

Specific examples of the form of the a_{kj} and b_k coefficients in (4.2) are provided for a lower bound on head and for a head difference constraint. For the lower bound on head all α_i^h are set to zero except the ith coefficient, which is set to one. The constant term, β^h, is set to the negative of the lower bound. This yields the constraint

$$h_i \geq h_i^l \tag{4.9}$$

which when combined with (4.4) yields

$$\sum_{j=1}^{n} \frac{\partial h_i}{\partial q_j} q_j \geq h_i^l - h_i^0 \tag{4.10}$$

The coefficients in (4.2) can be readily identified as

$$a_{kj} = \frac{\partial h_i}{\partial q_j}; \quad b_k = h_i^l - h_i^0 \tag{4.11}$$

For the head difference constraint, all α_i^h are set to zero with the exception of the coefficient on the first head in the head difference pair, which is set to one, and the second head in the pair, which is set to negative one. The constant term, β^h, is set to the negative of the head difference constraint. This results in a constraint of the form

$$h_{k_1} - h_{k_2} \geq h_k^d \tag{4.12}$$

which, after combination with (4.4), can be written as

$$\sum_{j=1}^{n} \frac{\partial h_{k_1}}{\partial q_j} q_j - \sum_{j=1}^{n} \frac{\partial h_{k_2}}{\partial q_j} q_j \geq h_k^d - \left(h_{k_1}^0 - h_{k_2}^0 \right) \tag{4.13}$$

This can be rearranged to form

$$\sum_{j=1}^{n} \left(\frac{\partial h_{k_1}}{\partial q_j} - \frac{\partial h_{k_2}}{\partial q_j} \right) q_j \geq h_k^d - \left(h_{k_1}^0 - h_{k_2}^0 \right) \tag{4.14}$$

The coefficients can then be identified as

$$a_{kj} = \frac{\partial h_{k_1}}{\partial q_j} - \frac{\partial h_{k_2}}{\partial q_j} = \frac{\partial \left(h_{k_1} - h_{k_2} \right)}{\partial q_j}; \quad b_k = h_k^d - \left(h_{k_1}^0 - h_{k_2}^0 \right) \quad (4.15)$$

In a similar fashion, any terms in the objective function that involve hydraulic heads can be reformulated to be written as a linear combination of stresses.

4.2 ESTIMATING RESPONSE COEFFICIENTS

The constraint coefficients in the general form (4.2), a_{kj}, that contain derivatives of head provide information on the response of groundwater flow to changes in stress. These coefficients are referred to as response coefficients. The linkage between the simulation model and the optimization formulation is provided by these coefficients.

The most commonly used method for computing response coefficients is perturbation. This method approximates the derivative of head with respect to stress based on finite differences. The derivative of head with respect to stress is approximated by a forward difference as described in Section 2.3.1

$$\frac{\partial h_i}{\partial q_j} \approx \frac{\Delta h_i}{\Delta q_j} = \frac{h_i(\mathbf{q}_{\Delta j}) - h_i(\mathbf{q}_0)}{q_{\Delta j} - q_j^0} \quad (4.16)$$

Here $q_{\Delta j}$ is the perturbed stress at the jth well and $\mathbf{q}_{\Delta j}$ is a vector of stresses that differs from $\mathbf{q}_0$ only in the jth element by an amount $(q_{\Delta j} - q_j^0)$.

As in any application of finite differences, the impact of the perturbation size on the precision of the derivative approximation should be considered. Finite difference theory shows that the error for a forward difference is proportional to the perturbation size. However, when the aquifer is modeled as linear, then the first truncated term in the Taylor series expansion (the second derivative of head with respect to stress) is zero and the first derivative is a constant. Computation of this derivative by finite differences will be independent of the perturbation size with precision limited only by the number of digits carried in the computation. The nonlinear response of head to stress characterized by aquifers modeled as unconfined will be discussed in detail in Chapter 8.

The relationship between perturbation size and precision in the derivative is related to round-off error when the difference in the numerator is taken. If the two computed heads in (4.16) are very close in value, then significant precision can be lost. The issue becomes more acute when considering the use of perturbation for calculating the response coefficient for head difference constraints. These response coefficients are approximated as

$$\frac{\partial(h_{k_1} - h_{k_2})}{\partial q_j} \approx \frac{\Delta(h_{k_1} - h_{k_2})}{\Delta q_j} \tag{4.17}$$

$$= \frac{\left(h_{k_1}(\mathbf{q}_{\Delta j}) - h_{k_2}(\mathbf{q}_{\Delta j})\right) - \left(h_{k_1}(\mathbf{q}_0) - h_{k_2}(\mathbf{q}_0)\right)}{q_{\Delta j} - q_{0j}}$$

This calculation involves the difference of a difference in the numerator. First, the difference in head between nearby locations, k_1 and k_2, may be small, thereby leading to loss of precision during the algebraic operation. Second, the difference in head difference may also be small. Hence, the possibility of significant round-off error is a concern.

EXAMPLE OF PERTURBATION PRECISION

Consider as an example a hypothetical square aquifer with a transmissivity of 5000 m^2/day and assume that a perturbation value of 200 m^3/day is used to determine the first derivative of head with respect to pumping. For this case, the following head values are obtained:

$$\begin{array}{ll} h_{k_1,0} & 35.11888 \\ h_{k_2,0} & 34.83335 \\ h_{k_1,\Delta j} & 35.10323 \\ h_{k_2,\Delta j} & 34.82014 \end{array}$$

where seven digits of precision are recorded. Computing the response coefficient for these values using (4.17) yields 0.0000122. Hence, seven digits of precision in the original heads have been reduced to three digits of precision in the response coefficient. If a smaller perturbation value is used or fewer digits are available in the head values, it is conceivable that the resulting response coefficients would have no significant digits.

The perturbation value should be sufficiently large to avoid round-off problems in the difference computation. The perturbation should not be so large as to induce any nonlinear responses in the modeled aquifer.

The iterative matrix solvers used to solve the discretized groundwater flow equation in some computer codes may introduce additional precision issues. These solvers produce heads that are precise only to the degree dictated by the user-specified convergence criteria. The user must ensure that sufficient digits are available to determine adequate precision in response coefficients.

4.3 SIMPLEX METHOD

We now turn our attention to solving the linear groundwater flow management problem. We first develop some intuition about the geometric characteristics of linear programming problems and then present a powerful, widely used solution technique called the simplex method.

A general linear form of the groundwater flow management problem was presented in (4.2) in scalar notation. For the purpose of developing the solution technique, it is easier to express the problem in equivalent vector form as

$$\text{minimize} \quad f = \mathbf{c}^t \mathbf{q} \tag{4.18}$$

$$\text{such that} \quad \mathbf{A}\mathbf{q} \geq \mathbf{b}$$

$$\mathbf{q}^l \leq \mathbf{q} \leq \mathbf{q}^u$$

where bold lowercase variables are column vectors, bold uppercase variables are matrices, and the superscript t is the transpose. Note that the stress bounds have been separated from the other constraints.

4.3.1 GEOMETRIC INTERPRETATION

Intuition about linear programming problems can be gained by first studying a simple problem with only two decision variables. Recall the water supply example developed in Section 3.7, where we seek to maximize water supply pumping subject to three drawdown constraints to prevent river losses to the aquifer. The water supply formulation is repeated here with the constraints restated in terms of the decision variables

$$\text{maximize} \quad f = \mathbf{c}^t \mathbf{q} \tag{4.19}$$

$$\text{such that} \quad \mathbf{A}\mathbf{q} \geq \mathbf{b}$$

$$\mathbf{q} \geq \mathbf{0}$$

With only two potential pumping wells, the linear problem can be graphically depicted as shown in Figure 4.1, where each axis represents one of the decision variables. Although the constraints are inequality equations, they can be shown graphically by drawing a line that represents each constraint as an equality. These lines then mark the boundary between solutions that satisfy the constraints and those that violate the design requirements. The three drawdown, or functional, constraints are shown by solid lines in Figure 4.1, and the nonnegativity constraints on $\mathbf{q}$ restrict the search for the solution to the first quadrant. The coefficients for the constraint equations plotted here are computed in Section 4.4.1. The set of points that satisfy all of the constraints is shaded and is termed the feasible region. The

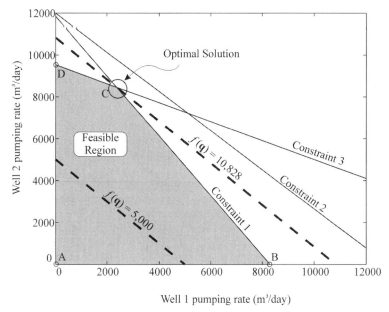

FIGURE 4.1   Graphical representation of the water supply example problem.

points where constraint boundaries intersect are called corner point solutions. The four feasible corner point solutions for the water supply problem are labeled in Figure 4.1 as points A, B, C, and D. Because the objective function and constraints are all linear, the optimal solution is guaranteed to be either a corner point solution or a line segment joining two adjacent corner points. In most cases, the optimal solution will be a point; however, it is possible that all points on a line segment may be optimal. This latter outcome is an example of multiple optima, which are discussed in Section 3.2.5.

The value of the objective function is the sum of the two pumping rates q_1 and q_2. Two contours of $f(\mathbf{q})$ are shown in Figure 4.1 as dashed lines. Imagine moving one of the dashed lines up and down the graph. As the line is moved toward increased pumping rates, the value of the objective function also increases. The objective function can increase until the boundary of the feasible region is reached, whereupon the optimal solution is found. Any further increase in pumping will result in violation of at least one of the constraints; in the context of the example problem, constraint violation implies that water from the river flows into the aquifer.

The optimal solution can be determined from Figure 4.1 as $q_1 = 2393$ m^3/day and $q_2 = 8435$ m^3/day for a total water supply of 10,828 m^3/day. Notice that two of the constraints are satisfied as equalities at the optimal solution. These two constraints, numbered 1 and 3, are called binding constraints because they prevent additional pumping and thus bind the solution. Conversely, constraint 2 is met as an inequality and is a nonbinding constraint at the solution.

In the preceding discussion, several characteristics of linear programming problems were identified in the context of the example problem. Although no formal proof is given here, these properties are generally true and can be summarized as follows:

    1. The intersection of feasible solutions to all constraints defines the feasible region. If the intersection is a null space, there is no solution to the problem. If the feasible region is not bounded, then the problem may not be solvable. Examples of infeasible and unbounded formulations are given in Chapter 3.

    2. If there is a solution to the linear programming problem, it will either be a corner point or all points that lie on a line segment joining two adjacent corner points. For problems that have more than two decision variables, the solution will either be a corner point or all points on a surface joining multiple adjacent corner points.

    3. There is a finite number of corner points, a subset of which are feasible.

    4. If the objective function value at a corner point is better than the value at all adjacent corner points, then there are no better points anywhere and the optimal solution has been found.

Properties 2 and 3 imply that the search for the solution can be restricted to a finite number of corner points. Therefore, the solution can be found in a finite number of steps. These characteristics of linear programming problems suggest an algorithm in which we seek to find a sequence of adjacent feasible corner points for which the value of the objective function improves. When improvement is no longer possible, 4 implies that the optimal solution has been found.

4.3.2 MATHEMATICS OF THE SIMPLEX METHOD

Although graphical solutions are possible for two-dimensional problems, an algebraic algorithm is needed to solve larger problems. In order to define the algorithmic rules, a convention must be adopted for the type of formulation to be solved. In this presentation, we assume that the objective function is to be maximized. Recall that a minimization problem can be transformed into a maximization problem by multiplying the objective function by (-1). Therefore, the derivations and procedures presented in the following are applicable to both minimization and maximization problems. Another assumption in the presentation to follow is that the decision variables either have only lower bounds of zero or, if present, the upper bound constraints are within the general constraint set. There are rules for including upper bounds in a manner similar to the nonnegativity constraints; however, for simplicity of presentation, these rules are not presented here.

As indicated before, we need to define mathematical rules to determine a sequence of adjacent feasible corner point solutions $\mathbf{q}^1, \mathbf{q}^2, \ldots, \mathbf{q}^*$ for which

$f(\mathbf{q}^1) < f(\mathbf{q}^2) < \cdots < f(\mathbf{q}^*)$ where $\mathbf{q}$ is the vector of decision variables and $\mathbf{q}^*$ is the optimal solution. Prior to presenting the simplex method, let us fully define the variables used to express the problem. The linear program is repeated here as

$$\text{maximize} \quad f = \mathbf{c}^t \mathbf{q} \qquad (4.20)$$

$$\text{such that} \quad \mathbf{A}\,\mathbf{q} \geq \mathbf{b}$$

$$\mathbf{q} \geq \mathbf{0}$$

where there are n decision variables and m functional constraints such that $\mathbf{c}$ is an $(n \times 1)$ vector of cost coefficients, $\mathbf{q}$ is the $(n \times 1)$ vector of decision variables, $\mathbf{A}$ is an $(m \times n)$ matrix of constraint coefficients, and $\mathbf{b}$ is the $(m \times 1)$ vector of right-hand side values for the functional constraints.

Overview of the Simplex Method

The simplex method is an iterative technique that proceeds by numerically finding adjacent feasible corner point solutions. At each corner point solution, the local sensitivity of the objective function to the decision variables is checked and the solution is moved to an adjacent corner point if improvement in the objective is possible. For the water supply example depicted in Figure 4.1, this process might include starting at location A, then proceeding to point B, and finally to C. An alternative path to the optimal solution is from A to D to C.

The objective function and constraints form a system of linear equations that can be manipulated and solved using the rules of matrix algebra. However, it is difficult to perform matrix operations on inequality equations so the simplex method adds one new variable to each functional constraint to transform the inequality equations into equalities. The augmented formulation then consists of m constraint equations in $(n + m)$ unknowns. Such a system of equations is indeterminate, but if n variables are set to fixed values then the remaining m variables can be uniquely determined. In the simplex method, the n variables that are set to fixed values are called nonbasic variables and are set to either their lower or upper bounds. In the example problem, the pumping rates have only lower bounds so the nonbasic variables ($n = 2$ in the example) are set to zero. The simplex method proceeds by systematically allowing one of the nonbasic variables to become nonzero while requiring one of the m nonzero, or basic, variables to go to zero. As shown in the next subsection, these steps are equivalent to finding adjacent corner point solutions.

Once the nonbasic variables are determined, the remaining system of m equations in m unknowns is solved by Gaussian elimination. The sensitivity of the objective function to a unit change in each nonbasic variable is then calculated at the current solution. In the simplex method, the nonbasic variable that would result in the largest marginal increase in the objective function is chosen to increase in value at the next iteration. Meanwhile, the constraints are used to limit the amount of feasible increase and to determine which of the basic variables goes to zero.

The process of choosing the n nonbasic variables and solving for the remaining m variables is done iteratively until no further improvement in the objective function value is possible. If all values representing the change in objective function value to a change in a nonbasic variable are negative, then increasing any of these variables above zero would result in deterioration of the objective function. When this situation occurs, no further improvement is possible and the optimal solution has been found.

In many problems, finding an initial feasible solution to begin the search is not a trivial problem. For simplicity we will assume an initial feasible solution is known. Those interested in learning more about procedures for starting the simplex method should consult an introductory linear progamming textbook, several of which are referenced at the end of this chapter.

Augmenting the Original Problem

Before a linear program can be solved algebraically, the inequality constraints must be converted into equality constraints. Transformation of the inequality constraints is accomplished by augmenting each constraint with a new variable; a slack variable is added to constraints of the "less than or equal to" form, and a surplus variable is subtracted from constraints of the form in the water supply example. Despite the nomenclature differences, we will use the term slack variable to include surplus variables. Upon augmenting the original problem, the new linear program becomes

$$\text{maximize} \quad f = \mathbf{c}^t \mathbf{q} + \mathbf{0}^t \mathbf{x}_s \tag{4.21}$$

$$\text{such that} \quad \mathbf{A}\,\mathbf{q} - \mathbf{I}\,\mathbf{x}_s = \mathbf{b}$$

$$\mathbf{q} \geq \mathbf{0}; \quad \mathbf{x}_s \geq \mathbf{0}$$

where $\mathbf{x}_s$ are slack variables and $\mathbf{0}$ is a vector of zeros. The vector $\mathbf{x}_s$ is an $(m \times 1)$ vector; notice that each slack variable is present in only one constraint equation. The formulation in (4.21) can be written compactly as

$$\text{maximize} \quad f = \underset{\sim}{\mathbf{c}}^t \mathbf{x} \tag{4.22}$$

$$\text{such that} \quad \underset{\sim}{\mathbf{A}}\mathbf{x} = \mathbf{b}$$

$$\mathbf{x} \geq \mathbf{0}$$

where the vectors $\underset{\sim}{\mathbf{c}} = \begin{bmatrix} \mathbf{c} \\ \mathbf{0} \end{bmatrix}$ and $\mathbf{x} = \begin{bmatrix} \mathbf{q} \\ \mathbf{x}_s \end{bmatrix}$ and the matrix $\underset{\sim}{\mathbf{A}} = [\mathbf{A} \mid -\mathbf{I}]$.

If a constraint is binding, the slack variable equals zero. Conversely, a positive value for a slack variable implies that the constraint is not binding at the solution. Recall the water supply example depicted in Figure 4.1. In the previous section, we determined that constraints 1 and 3 were met as equalities at the optimal solution. In order for the augmented constraints in (4.21) to be satisfied, the slack variables for these two constraints must both equal zero. Conversely, constraint 2 was met as an inequality at the optimal solution and the slack variable for this constraint must be nonzero for (4.21) to hold.

Staying with the example problem, we see in Figure 4.1 that there are four feasible corner point solutions, which are labeled A–D. With the augmented problem as stated in (4.21), the values of the variables q_1, q_2, x_{s1}, x_{s2}, x_{s3} at the points A–D are

$$
\begin{array}{llllll}
\text{A:} & (\quad 0 & 0 & 0.42 & 0.51 & 0.68\ ) \\
\text{B:} & (\ 8317 & 0\ \ 0 & & 0.18 & 0.41\ ) \\
\text{C:} & (\ 2393 & 8435 & 0 & 0.056 & 0\quad) \\
\text{D:} & (\quad 0 & 9532 & 0.082 & 0.010 & 0\quad)
\end{array}
$$

Notice that for each corner point, two variables equal zero and three variables have positive values. Also note that for adjacent corner points (e.g., A and B) two of the variables that have nonzero values are the same (e.g., x_{s2}, x_{s3}). The property of adjacent corner point solutions differing by only one nonzero, or basic, variable is a general property of linear programs that is exploited by the simplex method.

Basic and Nonbasic Variables

The addition of slack variables increases the number of problem variables to $(n + m)$ where there are n original decision variables and m slack variables. The problem now consists of m equations in $(n + m)$ unknowns. The inclusion of slacks for the example problem results in three constraint equations with five unknowns. In order to solve such a system of equations, n variables must be set to fixed values prior to solving for the remaining m variables. By setting the n fixed variables to zero the nonnegativity constraints are automatically met. These n fixed variables are called nonbasic variables and the others are called basic variables.

It is convenient to reorganize the vectors and matrices in (4.21) in terms of basic and nonbasic variables, $\mathbf{x}_B$ and $\mathbf{x}_N$, and their components such that

$$
\underline{\mathbf{c}} = \left\{ \begin{array}{c} \mathbf{c} \\ \mathbf{0} \end{array} \right\} = \left\{ \begin{array}{c} \mathbf{c}_B \\ \mathbf{c}_N \end{array} \right\} \tag{4.23}
$$

$$
\mathbf{x} = \left\{ \begin{array}{c} \mathbf{q} \\ \mathbf{x}_S \end{array} \right\} = \left\{ \begin{array}{c} \mathbf{x}_B \\ \mathbf{x}_N \end{array} \right\}
$$

$$
\underline{\mathbf{A}} = [\mathbf{A} \mid -\mathbf{I}] = [\mathbf{B} \mid \mathbf{N}]
$$

where $\mathbf{c}_B$ and $\mathbf{c}_N$ are the respective cost coefficients and $\mathbf{B}$ and $\mathbf{N}$ are matrices of constraint coefficients on basic and nonbasic variables, respectively.

The problem as stated in (4.21) can now be restated in terms of basic and nonbasic variables as

$$\text{maximize} \quad f = \mathbf{c}_B^t \mathbf{x}_B + \mathbf{c}_N^t \mathbf{x}_N \qquad (4.24)$$

$$\text{such that} \quad \mathbf{B}\mathbf{x}_B + \mathbf{N}\mathbf{x}_N = \mathbf{b}$$

$$\mathbf{x}_N = \mathbf{0}$$

where $\mathbf{B}$ is an $(m \times m)$ basis matrix with elements equal to the original constraint coefficients corresponding to the basic variables. Similarly, $\mathbf{N}$ is the $(m \times n)$ matrix of original constraint coefficients corresponding to the nonbasic variables. Note that because the simplex method sets nonbasic variables equal to fixed values, the values of the basic variables can be readily determined as

$$\mathbf{x}_B = \mathbf{B}^{-1}\left(\mathbf{b} - \mathbf{N}\mathbf{x}_N\right) \qquad (4.25)$$

Substituting (4.25) into the objective function, the updated function is

$$f = \mathbf{c}_B^t\left[\mathbf{B}^{-1}(\mathbf{b} - \mathbf{N}\mathbf{x}_N)\right] + \mathbf{c}_N^t\mathbf{x}_N \qquad (4.26)$$

or

$$f = \mathbf{c}_B^t\mathbf{B}^{-1}\mathbf{b} + \left(\mathbf{c}_N^t - \mathbf{c}_B^t\mathbf{B}^{-1}\mathbf{N}\right)\mathbf{x}_N \qquad (4.27)$$

Equation (4.27) can be further generalized by adding a term equal to zero to the right side

$$f = \mathbf{c}_B^t\mathbf{B}^{-1}\mathbf{b} + \left(\mathbf{c}_N^t - \mathbf{c}_B^t\mathbf{B}^{-1}\mathbf{N}\right)\mathbf{x}_N + \left(\mathbf{c}_B^t - \mathbf{c}_B^t\mathbf{B}^{-1}\mathbf{B}\right)\mathbf{x}_B \qquad (4.28)$$

and rearranging, using the relationships identified in equations (4.22) and (4.23)

$$f = \mathbf{c}_B^t\mathbf{B}^{-1}\mathbf{b} + \left(\underset{\sim}{\mathbf{c}}^t - \mathbf{c}_B^t\mathbf{B}^{-1}\underset{\sim}{\mathbf{A}}\right)\mathbf{x} \qquad (4.29)$$

The coefficients on $\mathbf{x}$ in (4.29) are called the reduced costs and give the marginal improvement in f for a small change in each variable. Note that the second term in (4.29) is zero because the nonbasic variables themselves are zero and the coefficients on the basics are zero. Nonetheless, both terms on the right side of (4.29) are retained because the first term is the objective function value at the current solution and the second term provides information on how to improve upon the current solution.

As stated previously, the goal of the simplex method is to move from one feasible corner point to an adjacent feasible corner point that has a better objective function value. Adjacent corner points differ by only one basic variable, so to move to an adjacent point one basic variable must leave the basis (the leaving basic variable) and become nonbasic while one nonbasic variable must become basic (the entering basic variable). The next two sections describe the methods and logic used to find adjacent feasible corner point solutions and ultimately the optimal solution.

Finding the Entering Basic Variable

In the previous subsection, we saw that the objective function value for any set of basic and nonbasic variables can be determined from (4.29). The reduced costs represent the local sensitivity of f to small changes in each variable. If the reduced cost of a nonbasic variable is positive, then increasing the value of that variable will result in an increase in the objective function. Recall that the problem is to maximize f; therefore any positive reduced cost for a nonbasic variable implies that the objective function can be increased further.

What if several coefficients are positive? Although there is no way of knowing *a priori* which nonbasic variable we should choose to become the entering basic variable, the simplex method chooses the variable with the largest positive reduced cost. This results in the largest marginal increase in the objective function value.

When a set of basic and nonbasic variables is such that all reduced costs are nonpositive, then no further improvement in f is possible and the optimal solution has been found.

Finding the Leaving Basic Variable

While we have just shown how to determine the entering basic variable, we also recognize that the entering variable cannot be increased indefinitely. At some point, the solution will become infeasible and the constraints are used to control the amount of increase in the entering variable. One way of determining this limit is to perform Gaussian elimination on the constraint matrix. If all but one basic variable is removed from each constraint, then a simple relationship between each basic variable and all of the nonbasic variables is developed. This step can be performed by premultiplying both sides of the constraints in (4.24) by $\mathbf{B}^{-1}$

$$\mathbf{B}^{-1}\left(\mathbf{B}\,\mathbf{x}_B + \mathbf{N}\,\mathbf{x}_N\right) = \mathbf{B}^{-1}\,\mathbf{b} \tag{4.30}$$

and rearranging

$$\mathbf{I}\,\mathbf{x}_B + \left(\mathbf{B}^{-1}\,\mathbf{N}\right)\mathbf{x}_N = \mathbf{B}^{-1}\,\mathbf{b} \tag{4.31}$$

Notice that each basic variable has a coefficient of one in only one constraint and a coefficient of zero in all other constraints. Also note that all current coefficients are obtained using information from the original problem definition.

To determine the relationship between the entering basic variable and a current basic variable, let x_j be the entering variable and look at the constraint for basic variable x_i

$$x_i + \left(\mathbf{B}^{-1}\,\mathbf{N}\right)_{ij} x_j = \left(\mathbf{B}^{-1}\,\mathbf{b}\right)_i \tag{4.32}$$

where $(\mathbf{B}^{-1}\mathbf{N})_{ij}$ is the element in the matrix $(\mathbf{B}^{-1}\mathbf{N})$ corresponding to variable x_j in row i and $(\mathbf{B}^{-1}\mathbf{b})_i$ is the current value of basic variable x_i. Prior to changing the value of either variable, we know $x_i = (\mathbf{B}^{-1}\mathbf{b})_i$ because all nonbasic variables are zero. If x_i is to be removed from the basis, then it will be reduced to zero and x_j will increase to $(\mathbf{B}^{-1}\mathbf{b})_i/(\mathbf{B}^{-1}\mathbf{N})_{ij}$. If (4.32) is considered for each constraint equation, the smallest ratio provides the limiting value on the feasible increase in x_j. The x_i associated with the limiting ratio goes to zero and thus becomes nonbasic. Therefore, the leaving basic variable is determined by the ratio test

$$\min_i \left\{ \frac{(\mathbf{B}^{-1}\mathbf{b})_i}{(\mathbf{B}^{-1}\mathbf{N})_{ij}} \,\middle|\, (\mathbf{B}^{-1}\mathbf{N})_{ij} > 0 \right\} \qquad (4.33)$$

In summary, the simplex method sets n variables to fixed values and solves for the remaining m variables by Gaussian elimination at each iteration. The reduced costs are used to determine the entering variable x_j, and the constraints are used to determine the leaving variable x_i. The basis matrix $\mathbf{B}$ is updated at each iteration by simply replacing the column associated with x_i with the original coefficients associated with x_j. The reduced costs are then recalculated per (4.29) and the preceding steps are repeated until all reduced costs are nonpositive and no further improvement in f is possible. Equations (4.25) and (4.29) are then solved to determine the optimal solution.

4.4 SOLVING THE WATER SUPPLY EXAMPLE

In this section, we continue the water supply example introduced in Chapter 3 and use it to demonstrate the solution method elements introduced in this chapter. The values of the response coefficients in the water supply example are determined and the problem is solved using the simplex method as outlined in the previous section. In scalar notation, the water supply example is

$$\text{maximize} \quad q_1 + q_2 \qquad (4.34)$$

$$\text{such that}$$

$$h_A \geq h_{\text{river}, A}$$

$$h_B \geq h_{\text{river}, B}$$

$$h_C \geq h_{\text{river}, C}$$

$$q_1 \geq 0, \quad q_2 \geq 0$$

4.4.1 COMPUTE RESPONSE COEFFICIENTS

Substituting the Taylor series representation for the constraint statements yields

$$\text{maximize} \quad q_1 + q_2 \tag{4.35}$$

such that

$$\frac{\partial h_A}{\partial q_1} q_1 + \frac{\partial h_A}{\partial q_2} q_2 \geq h_{\text{river}, A} - h_A^0$$

$$\frac{\partial h_B}{\partial q_1} q_1 + \frac{\partial h_B}{\partial q_2} q_2 \geq h_{\text{river}, B} - h_B^0$$

$$\frac{\partial h_C}{\partial q_1} q_1 + \frac{\partial h_C}{\partial q_2} q_2 \geq h_{\text{river}, C} - h_C^0$$

$$q_1 \geq 0, \quad q_2 \geq 0$$

In order to solve this formulation, values for the response coefficients and right-hand side values for all three constraints must be determined. A numerical model of this problem was developed and used to determine the relevant values. The values for the right side of the constraints are determined with information calculated in the initial groundwater simulation:

Location, j	$h_{\text{river}, j}$	h_j^0	$h_{\text{river}, j} - h_j^0$
A	504.9	505.319177	−0.41918
B	504.7	505.205849	−0.50585
C	504.3	504.977716	−0.67772

The constraint coefficients must be determined by numerical perturbation as described in Section 4.2. Based on a perturbation value of $\Delta q = 100 \text{ m}^3/\text{day}$, relevant values for the water supply example are as follows:

Head–well pair	Head base value	Head perturbed value	Response coefficient
A, 1	505.319177	505.314138	-5.04×10^{-5}
A, 2		505.315633	-3.54×10^{-5}
B, 1	505.205849	505.201907	-3.94×10^{-5}
B, 2		505.201631	-4.22×10^{-5}
C, 1	504.977716	504.974459	-3.26×10^{-5}
C, 2		504.970603	-7.11×10^{-5}

4.4.2 APPLY SIMPLEX METHOD

With the information just provided, the water supply problem can now be solved by the simplex method. In vector notation, the augmented problem is stated as

$$\text{maximize} \quad f = \underset{\sim}{c}{}^t x \tag{4.36}$$

$$\text{such that} \quad \underset{\sim}{A} x = b$$

$$x \geq 0$$

where $\underset{\sim}{c}$, x, and $\underset{\sim}{A}$ are understood to include the slack variables and their coefficients. Substituting the numerical values determined for the vector and matrix elements,

$$\underset{\sim}{c}{}^t = [\,1 \quad 1 \quad 0 \quad 0 \quad 0\,]$$

$$x^t = [\,q_1 \quad q_2 \quad x_{s1} \quad x_{s2} \quad x_{s3}\,]$$

$$\underset{\sim}{A} = \begin{bmatrix} -0.0000504 & -0.0000354 & -1 & 0 & 0 \\ -0.0000394 & -0.0000422 & 0 & -1 & 0 \\ -0.0000326 & -0.0000711 & 0 & 0 & -1 \end{bmatrix} \tag{4.37}$$

$$b^t = \begin{bmatrix} -0.41918 & -0.50585 & -0.67772 \end{bmatrix}$$

In this simple example, we know that the constraints are satisfied when pumping is not occurring from either of the supply wells (corner point A in Figure 4.1). Using the two supply wells as the initial nonbasic variables, the first simplex iteration would have the following list of basic and nonbasic variables

$$x_B^t = [\,x_{s1} \quad x_{s2} \quad x_{s3}\,]$$

$$x_N^t = [\,q_1 \quad q_2\,] = [\,0 \quad 0\,] \tag{4.38}$$

Recall that the basis matrix B and the matrix N are derived as the columns of the original coefficient matrix $\underset{\sim}{A}$ corresponding to the basic and nonbasic variables, respectively. For the first iteration,

$$B = \begin{bmatrix} -1 & 0 & 0 \\ 0 & -1 & 0 \\ 0 & 0 & -1 \end{bmatrix} ; \quad N = \begin{bmatrix} -0.0000504 & -0.0000354 \\ -0.0000394 & -0.0000422 \\ -0.0000326 & -0.0000711 \end{bmatrix} \tag{4.39}$$

From (4.25) and (4.29) the values of the basic variables and the reduced costs at the first iteration are determined as

$$x_B^t = \begin{bmatrix} B^{-1}(b - Nx_N) \end{bmatrix}^t = [\,0.4192 \quad 0.5059 \quad 0.6777\,] \tag{4.40}$$

$$\left(\underset{\sim}{c}{}^t - c_B^t B^{-1} \underset{\sim}{A} \right) = [\,1 \quad 1 \quad 0 \quad 0 \quad 0\,] \tag{4.41}$$

Although (4.40) gives us the value of the basic variables at the first iteration, this information is not used in deciding the next step. Instead, it is the values in (4.41) that are used to determine how to alter the basis and in the process move to an adjacent corner point solution. The two nonbasic variables (q_1, q_2) both have positive reduced costs in (4.41). This means that the objective function can be increased if either of these variables becomes positive. Because the coefficients are equal, the marginal improvement is the same regardless of which nonbasic variable is chosen to become basic. In the event of a tie such as seen here, the entering basic variable is chosen randomly. Let's allow the nonbasic variable q_1 to become the entering basic variable.

To determine the leaving basic variable, we must look at the constraints because they restrict the amount of feasible increase for q_1. From (4.33), the leaving basic variable is determined as the one that provides the limiting ratio

$$\frac{1}{(\mathbf{B}^{-1}\mathbf{N})_{i1}}(\mathbf{B}^{-1}\mathbf{b})_i = \begin{bmatrix} 8317 \\ 12839 \\ 20789 \end{bmatrix} \tag{4.42}$$

These results tell us that the first constraint limits the amount of allowable increase in variable q_1. In fact, q_1 can increase to 8317 m³/day, at which point constraint 1 becomes binding (see corner point B in Figure 4.1). When a constraint is binding, the slack variable for that constraint must equal zero. Because slack variable x_{s1} goes to zero, it leaves the basis and becomes nonbasic.

The second iteration begins by replacing q_1 for x_{s1} in the basis. Vectors and matrices for the second iteration are

$$\mathbf{x}_B^t = \begin{bmatrix} q_1 & x_{s2} & x_{s3} \end{bmatrix}$$
$$\mathbf{x}_N^t = \begin{bmatrix} x_{s1} & q_2 \end{bmatrix} = \begin{bmatrix} 0 & 0 \end{bmatrix} \tag{4.43}$$

$$\mathbf{B} = \begin{bmatrix} -0.0000504 & 0 & 0 \\ -0.0000394 & -1 & 0 \\ -0.0000326 & 0 & -1 \end{bmatrix}; \quad \mathbf{N} = \begin{bmatrix} -1 & -0.0000354 \\ 0 & -0.0000422 \\ 0 & -0.0000711 \end{bmatrix}$$

As before, we now determine the values of the basic variables and calculate the new reduced costs.

$$\mathbf{x}_B^t = \begin{bmatrix} \mathbf{B}^{-1}(\mathbf{b} - \mathbf{N}\mathbf{x}_N) \end{bmatrix}^t = \begin{bmatrix} 8317 & 0.1782 & 0.4066 \end{bmatrix} \tag{4.44}$$

$$\left(\underset{\sim}{\mathbf{c}}^t - \mathbf{c}_B^t \mathbf{B}^{-1}\underset{\sim}{\mathbf{A}} \right) = \begin{bmatrix} 0 & 0.2976 & -19841 & 0 & 0 \end{bmatrix} \tag{4.45}$$

In the second iteration only one reduced cost is positive and it corresponds to nonbasic variable q_2. The leaving basic variable is again determined as the variable that produces the limiting ratio in the constraint equations

$$\frac{1}{(\mathbf{B}^{-1}\mathbf{N})_{i2}}(\mathbf{B}^{-1}\mathbf{b})_i = \begin{bmatrix} 11841 \\ 12264 \\ 8435 \end{bmatrix} \tag{4.46}$$

This tells us that the third basic variable, in this case x_{s3}, is the leaving variable.

As before, the third iteration begins by replacing q_2 for x_{s3} in the basis. Vectors and matrices for the third iteration are

$$\mathbf{x}_B^t = [\, q_1 \quad x_{s2} \quad q_2 \,]$$
$$\mathbf{x}_N^t = [\, x_{s1} \quad x_{s3} \,] = [0 \quad 0] \tag{4.47}$$

$$\mathbf{B} = \begin{bmatrix} -0.0000504 & 0 & -0.0000354 \\ -0.0000394 & -1 & -0.0000422 \\ -0.0000326 & 0 & -0.0000711 \end{bmatrix} ; \quad \mathbf{N} = \begin{bmatrix} -1 & 0 \\ 0 & 0 \\ 0 & -1 \end{bmatrix}$$

Again, we now determine the values of the basic variables and update the reduced costs

$$\mathbf{x}_B^t = \left[\mathbf{B}^{-1}(\mathbf{b} - \mathbf{N}\mathbf{x}_N) \right]^t = [\, 2393 \quad 0.0556 \quad 8435 \,] \tag{4.48}$$

$$(\underline{c}^t - \mathbf{c}_B^t \mathbf{B}^{-1} \underline{\mathbf{A}}) = [0 \quad 0 \quad -15847 \quad 0 \quad -6174\,] \tag{4.49}$$

Because all reduced costs are now less than or equal to zero, the objective function cannot be improved any further. Thus, we have found the optimal solution, which is $q_1 = 2393$ m^3/day, $q_2 = 8435$ m^3/day, $x_{s1} = 0$, $x_{s2} = 0.0556$, and $x_{s3} = 0$.

4.5 ESTIMATING COMPUTATIONAL EFFORT

For the linear groundwater flow management problem, there are two computational issues to consider: calculating problem coefficients and solving the linear program. As discussed in the following, the effort required to determine response coefficients is related to the number of candidate wells whereas the effort in the solution step is related to the number of constraints. Combined, these facts point to the need to develop well-posed management formulations that do not include excessive numbers of candidate wells or constraints.

As presented in Section 4.2, constraint coefficients are determined by estimating response coefficients. This in turn requires a numerical differencing scheme where the groundwater simulator is called for each perturbation at each pumping well. The computational effort of the groundwater simulator will depend on the number of numerical grid cells and the solution technique employed to determine the head distribution. If the groundwater model has many numerical grid cells, each call to the simulator is computationally intensive. In all cases, the number of calls to the simulator equals one initial call to determine background heads plus one call per candidate well.

The primary computational burden in solving a linear program by the simplex method is calculating the inverse of the basis matrix. Recall from our example problem that the basis inverse was used to determine both the entering and leaving basic variables at each iteration. Fortunately, the updated basis matrix differs from the previous basis by only one column and efficient techniques have been developed to update the inverse basis matrix. Nonetheless, full matrix inversion is required periodically during the solution process and can be computationally intensive. Depending on the procedure used, computational effort is typically on the order of the number of rows in the matrix cubed. Thus, the computational effort needed to solve the linear program is more strongly influenced by the number of functional constraints than by the number of decision variables. Furthermore, the number of basis inversions is controlled by the number of simplex iterations, which is bounded by the number of corner points. For a system with n decision variables (including slacks) and m constraints, the number of corner point solutions is $\frac{(m+n)!}{m!n!}$, only a subset of which are feasible. Although the number of corner points can be quite large, experience suggests that the number of iterations to convergence is typically about two times the number of functional constraints.

4.6 NOTES AND REFERENCES

Response coefficients and response matrices, as described in Section 4.1, have been used since the first applications of optimization to groundwater flow problems (Deninger, 1970; Maddock, 1972). An alternative method for coupling the optimization and simulation models is to include the governing equation directly as constraints. In this approach, the system of numerically discretized algebraic equations that approximate the governing equation are included as constraints in the linear program. This method is known as the embedding approach and was first presented in a groundwater management model by Aguado and Remson (1974). Numerical stability issues can arise when using this method that have been attributed to poorly conditioned constraint matrices (Peralta *et al.*, 1991). Although methods exist for handling such problems, the response matrix approach is generally favored.

The numerical issues related to the calculation of response matrix coefficients described in Section 4.2 are explored by Riefler and Ahlfeld (1996). Alternatives to the perturbation approach for computing the response coefficients are available using adjoint operators (Sykes *et al.*, 1985). The adjoint approach may be computationally superior to the perturbation approach if the number of constraints that involve hydraulic head is smaller than the number of candidate stresses.

The simplex method was developed by Dantzig in the 1940s and 1950s (Dantzig, 1963). The method is now routinely applied to large problems in numerous industrial, economic, and management fields. Introductory linear programming textbooks include those by Bradley et al. (1977), Solow (1984), and Hillier and Lieberman (1995). More advanced numerical issues associated with computer implementation of the simplex method are addressed by Murtagh (1981) and Nazareth (1987).

5

USING THE
MANAGEMENT MODEL

This chapter presents several different techniques for investigating the relationships between problem formulation and optimal solutions. The set of decision variables, constraints, and parameter values for a specific formulation all interact to define the feasible space and the optimal solution. However, management models are often developed without complete information about the groundwater system, the design criteria, or policy constraints. It is therefore important to investigate the effect of variations in the set of decision variables, constraints, and parameter values on the optimal solution. Information about the impact of these variations can be gained through the processes of sensitivity and range analysis and by examination of the dual problem. These techniques lead to further insight into the nature of the management problem and provide the analyst and decision makers with additional tools for assessing alternatives.

5.1 SENSITIVITY ANALYSIS

When using simulation models it is common to examine the uncertainty in the results of the model by perturbing selected input parameters in a procedure known as sensitivity analysis. Parameters are varied within reasonable ranges and the model output is examined accordingly. A similar concept can be used to examine the sensitivity of the groundwater flow management model to its parameters.

Calibration and sensitivity analysis are closely related because they both involve examination of model results as a function of changes in model parameter values. In general, the parameters of the model, including the locations of the constraints and candidate wells, are values that can be modified in a sensitivity analysis. The choice of parameters to vary will generally be driven by policy considerations or cost issues.

When performing sensitivity analysis there are several ways to measure the impact of changing model parameters on the solution. Such measures include changes in the total stress, the number of stress locations, or the robustness of stress locations. It may be necessary to organize the sensitivity analysis to control changes in one or more of these measures.

A common scenario for sensitivity analysis is that a particular solution is determined and the sensitivity of stresses at the active locations is to be tested. If the original set of candidate locations is retained during sensitivity analysis, it is possible that new stress locations will be selected. This outcome is inconsistent with the goal of testing the active wells from the original formulation. One means of addressing this is simply to eliminate the candidate locations that are not selected in the original optimization. If this approach is chosen then it is implied that sensitivities are no longer being determined relative to the original optimization problem, but the sensitivity is relative to a subset of candidate locations.

5.1.1 NUMBER OR LOCATION OF STRESSES
OR CONSTRAINTS

Candidate Stresses

All of the optimization formulations presented in this book require that the set of candidate stress locations be fixed for a given formulation. An important issue to investigate is how the optimal stress rates and locations change under different sets of candidate locations. To perform this type of sensitivity analysis, alternative sets of candidate wells can be provided and the optimal solution determined for each set. Such an analysis can lead to insight into the trade-offs between the number of stress locations and total stress or can indicate the relative importance of specific stress locations in satisfying the constraints. Factors external to the management model may favor a design that is sub-optimal relative to the objective function value. Sensitivity analysis provides valuable information for making informed decisions and for understanding the trade-offs implied by certain designs.

EXAMPLE OF CHANGING THE SET OF
CANDIDATE STRESSES

Recall the construction dewatering example from Section 3.3.2, where hydraulic control must be obtained over a certain area and a solution has been achieved with four active wells from a set of nine candidate locations. Assume that for cost reasons it is desirable to

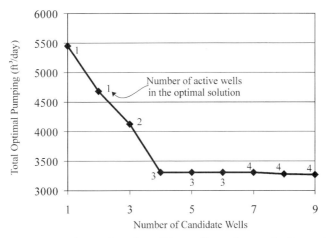

FIGURE 5.1 The number of candidate wells can have a significant effect on the optimal pumping scheme. These data are for the dewatering example depicted in Figure 3.7.

limit the number of wells for which piping has to be provided. Further assume that no stress locations other than the original nine can be considered for the application of stress. By using successive optimization with different sets of candidate locations, several solutions are found with pumping distributed between various wells. Complete enumeration of these solutions is impractical because there are 511 subsets of the original nine candidate wells. Instead, we can get an idea of solution behavior by checking just a few of these subsets. The results displayed in Figure 5.1 were generated by successively eliminating one candidate well at a time and re-solving the resulting formulation. The candidate wells were eliminated according to their identification numbers in Figure 3.7, so, for example, the formulation with seven candidate wells consists of wells numbered 1 through 7.

From Figure 5.1, we see that as the number of candidate locations decreases, the total pumping costs increase. This behavior results because some previously active wells are no longer available and pumping must increase at the remaining wells to effect the same level of control. Although four wells extracting at 3271 ft³/day result in the least amount of total pumping, Figure 5.1 indicates that a solution employing only three wells can satisfy the design constraints with a 1% increase in pumping. Factors external to the management model, such as construction costs or client preferences, may lead the decision maker to prefer the option with three wells and slightly more pumping. Similarly, Figure 5.1 shows that two solutions require only one active well, yet the pumping rates from the wells are quite different. Again, externalities may favor a "nonoptimal" design.

Constraints and Slack Variables

In most cases, only a subset of constraints will be met as equalities at the optimal solution. These constraints restrict the values of the decision variables and are said to bind the solution. Conversely, nonbinding constraints do not exert any influence over the optimal values of the decision variables and could be removed from the formulation without affecting the solution.

Information about which constraints bind the optimal solution is readily available from examination of the solution. This is particularly easy for linear problems solved with the simplex algorithm. Recall that slack variables are added to inequality constraints of the "less than or equal to" form and subtracted from constraints of the "greater than or equal to" form prior to solving the augmented formulation. Assuming that constraints are of the "less than or equal to" form, the augmented constraints are

$$\mathbf{Aq} + \mathbf{Ix}_s = \mathbf{b} \tag{5.1}$$

Rearranging (5.1) to solve for the slack variables,

$$\mathbf{x}_s = \mathbf{b} - \mathbf{Aq} \tag{5.2}$$

These variables can now be seen to represent the slack in each constraint, or the difference between the right-hand side value, b, and the constraint value. Therefore, slack variables for binding constraints have a value of zero at the optimal solution. When a slack variable is nonzero at the optimal solution, the value indicates either the amount of a resource not utilized by the optimal solution (the amount of slack in the constraint) or the excess in the constraint (the surplus).

Consider the water supply example depicted graphically in Figure 4.1. As previously noted, constraint 2 is nonbinding at the solution. If this constraint is removed from the formulation with no alterations to other parameters, the optimal solution does not change. Conversely, removal of constraints 1 or 3 from the formulation will allow considerably more pumping.

5.1.2 RIGHT-HAND SIDE CONSTRAINT VALUES

The general linear constraints in (3.23) contain two constant terms, β_k^q and β_k^h. When specific constraints are developed, such as the head difference constraint in (3.11) or the head bounds in (3.9) and (3.10), the constant terms are combined and moved to the right-hand side of each constraint equation. The set of specific constraint functions is expressed in vector notation as

$$\underset{\sim}{\mathbf{A}}\mathbf{x} = \mathbf{b} \tag{5.3}$$

where all the constant terms are contained in the $\mathbf{b}$ vector.

In many problems, the constraints act as surrogates for the true design goal. As such, the right-hand side values of the constraints are not well defined and may contain considerable uncertainty. In other management problems, the constraint might represent a policy goal, which could change if the situation is favorable. In either case, it is important to investigate the effect of different constraint values on the optimal solution. For example, we might ask how the optimal solution would change if the right-hand side of constraint i were changed by an amount Δb_i. The new constraint is stated as

$$\underset{\sim}{\mathbf{A}}_i \mathbf{x} = b_i + \Delta b_i \tag{5.4}$$

where $\underset{\sim}{\mathbf{A}}_i$ is the ith row of the $\underset{\sim}{\mathbf{A}}$ matrix. When using the simplex method, information concerning the effect of small unit changes in constraint values on the value of the objective is given by the shadow price. The effect of large perturbations in right-hand side values may require re-solving the linear program for each new value of the constraint. This process provides information to generate a trade-off curve between constraint values and the optimal solution.

Local Sensitivity and Shadow Prices

Local sensitivity values, also called shadow prices, represent the change in objective function value for a unit change in a specific right-hand side value. For example, the shadow prices for the water supply example indicate how total water supply would change for a unit change in each head constraint at the river nodes. This information is important for policy constraints because the marginal increase in benefit may be sufficient to consider allowing some relaxation of the constraint. Alternatively, for constraints in which there is considerable uncertainty in a right-hand side value, the shadow price can be used to determine whether additional resources should be devoted to reducing the uncertainty.

In mathematical notation, the shadow price for constraint i is determined from the optimal solution and the final updated objective function represented by (4.29) as

$$\frac{\partial f}{\partial b_i} = \frac{\partial}{\partial b_i} \left[\mathbf{c}_B^t \mathbf{B}^{-1} \mathbf{b} + \left(\underset{\sim}{\mathbf{c}}^t - \mathbf{c}_B^t \mathbf{B}^{-1} \underset{\sim}{\mathbf{A}} \right) \mathbf{x} \right] = \mathbf{c}_B^t \mathbf{B}_i^{-1} \tag{5.5}$$

where $\mathbf{B}_i^{-1}$ is the ith column of $\mathbf{B}^{-1}$. Note that the shadow prices are independent of the optimal decision variable values and are constant as long as the optimal basis remains the same.

To develop some intuition about shadow prices, first consider a nonbinding constraint. Because there is slack in the constraint, a small change in the right-hand side value will change the amount of slack but will have no effect on the objective function value. Therefore, the shadow price of a nonbinding constraint is zero. Conversely, binding constraints restrict the value of the decision variables and the objective function. Therefore, when the right hand side of a binding constraint increases by a unit amount, the objective function will change by an amount given by the shadow price as long as the optimal basis does not change. This implies that the optimal solution is quite sensitive to constraints that have large shadow price magnitudes.

EXAMPLE OF SHADOW PRICE

For the water supply example, constraints 1 and 3 have shadow prices of -15847 and -6174 (m^3/day/m), respectively. These shadow prices can be calculated using (5.5) and the information in Section 4.4. Alternatively, the shadow prices can be read directly from the optimal reduced costs in (4.49) by recognizing that the original objective function coefficient for slacks is zero and the original constraint coefficient for a slack is unity. The negative signs imply that the objective function will increase if the right-hand side is decreased. Thus, water supply pumping could be increased considerably if drawdown requirements at the river are relaxed, particularly at location 1. In this instance, it may be relevant for the water supply manager to study the implications of allowing some river flow to the aquifer.

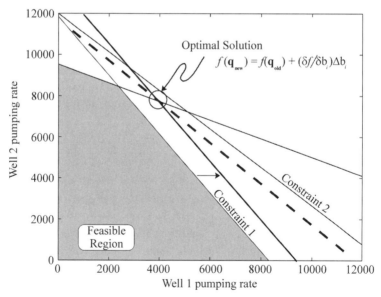

FIGURE 5.2   The shadow price indicates the amount the optimal solution will change for a unit change in a constraint value. Here, constraint 1 has been relaxed and the total pumping increases.

Consider relaxing the head constraint at location 1 by 0.05 m, so that the new head bound is 504.85 m rather than the original value of 504.9 m. The shadow price indicates that the total water supply will increase by $(-15{,}847 \text{ m}^3/\text{day/m}) \times (-0.05 \text{ m})$, which equals 792 m³/day. The geometric interpretation of the change in constraint is shown in Figure 5.2, where constraint 1 has shifted to the right, reflecting the new head bound. The new optimal solution is $q_1 = 3880$ m³/day and $q_2 = 7740$ m³/day, for a total increase in water supply of 792 m³/day.

To gain insight into the local nature of shadow prices, consider relaxing constraint 1 even further. In Figure 5.2, this is equivalent to shifting constraint 1 farther to the right. If constraint 1 shifts to the right of the intersection of constraints 2 and 3, then constraint 1 will no longer bind the solution and its shadow price will become zero. When this occurs, the basis changes and the shadow prices from the original formulation are no longer valid. Range analysis, presented in Section 5.2, determines the constraint values over which a shadow price is valid.

Trade-off Curves

While the shadow price measures marginal sensitivity, it is also useful to investigate the impact of larger changes in constraint values on the optimal solution and the objective function. Insight into the importance of a particular constraint can be gained by changing a constraint value over a large range and observing the resulting solution.

EXAMPLE OF TRADE-OFF CURVE

Continuing with the water supply example, we can deduce that constraint 2, which is nonbinding at the solution, can be relaxed infinitely without affecting the optimal solution. However, any changes in the values of constraints 1 or 3 will have an immediate impact on total pumping. Figure 5.3 depicts the effect of different right-hand side values for constraint 1 on the total optimal pumping. The active wells at each solution are also identified in the figure. Note that as the constraint on head in the aquifer is relaxed more pumping is allowed and the objective function increases. This curve shows the trade-off implied by changes in the policy related to river losses at location 1. Relaxing the constraint slightly from 504.9 m to 504.8 m increases total pumping by 14%, but a further 0.1 m decrease in the head bound increases supply by an additional 3%. When the constrained head falls below 504.6 m constraint 1 no longer binds the solution and water supply is unaffected by the constraint. Conversely, making the constraint more restrictive by increasing the head bound results in a smaller water supply and extraction shifts entirely to well 2.

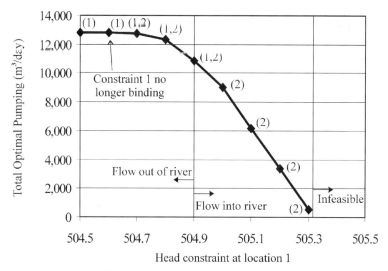

FIGURE 5.3   Effect of changes in contraint 1 on the optimal solution. The wells that are active in each solution are identified in parentheses.

5.1.3 MULTIPLE OBJECTIVES AND THE EFFICIENT FRONTIER

A critical component in formulating a management model is defining project objectives and constraints. In many, if not all, cases the management model can be expressed in several different ways. Furthermore, some problems may be more appropriately expressed with multiple objectives rather than a single objective. For example, a water supply problem might require maximizing water supply while minimizing ecological health risk to an adjacent wetland. As water supply extraction increases, flux to the wetland decreases and ecological health may suffer, indicating that there is a trade-off between the two objectives. As one objective improves (e.g., water supply increases), the other objective deteriorates (risk increases). Such problems inherently consist of trade-offs between the multiple objectives and do not have one optimal solution. Instead, there exists a set of solutions that are noninferior to each other, where each element of the set represents the optimal solution for a fixed trade-off preference level.

The set of noninferior solutions can be generated by representing different trade-off preferences numerically. This is accomplished by multiplying each objective by a coefficient, where the coefficient is a relative weighting of the different objectives. By varying the weights, the relative importance of the functions can be manipulated and the optimal solution will depend on the weight. Such an objective function for two functions can be stated as

$$\text{maximize } f(\mathbf{q}) = w[f_1(\mathbf{q})] - (1 - w)[f_2(\mathbf{q})] \tag{5.6}$$

where w is a weight. The set of optimal solutions over a range of weights results in a trade-off curve between the two objectives. This trade-off curve is alternatively called the efficient frontier or Pareto optimal solution. None of the solutions on the curve are better than the other solutions at meeting both objectives. In multiobjective programming, a decision process external to the optimization technique must be used to identify the efficient solution that best meets management goals.

EXAMPLE OF MULTIOBJECTIVE WATER SUPPLY

Consider again the water supply problem. In Section 3.7, two different formulations were proposed for expressing the need to have a water supply while simultaneously considering the impact of pumping on the nearby river. One formulation seeks to maximize extraction pumping while restricting groundwater heads at the river and the other formulation seeks to minimize river losses subject to meeting water supply demands.

A third formulation can be developed that combines the two objectives into one statement that requires maximizing water supply pumping while minimizing drawdown in the aquifer below the river. The problem will be unbounded unless we impose drawdown constraints, but a trade-off between total drawdown below the river and water supply is still implied by the objective function. This groundwater flow management model is expressed mathematically as

$$\text{maximize } f(\mathbf{q}) = w \sum_{i=1}^{2} q_i - (1 - w) \sum_{j=1}^{3} s_j \tag{5.7}$$

$$\text{such that} \qquad h_k \geq h_k^l \qquad k = 1, 2, 3 \tag{5.8}$$

$$\mathbf{q} \geq \mathbf{0} \tag{5.9}$$

where s_j is the drawdown at location j. Note that the competing objectives can be expressed in a combined statement because maximizing the negative of a function is equivalent to minimizing the original function.

The efficient frontier for this problem is depicted in Figure 5.4. The curve shown in the figure traces the path from corner point A to B to C in Figure 4.1. When minimizing drawdown is important, then Figure 5.4 indicates that all extraction should occur from well 1. The change in slope occurs when the lower head bound at constraint 1 is met and extraction commences at well 2. The upper limit of the curve identifies the maximum extraction that can occur from the two wells without violating the head bound constraints. Any additional extraction will result in flux from the river to the aquifer.

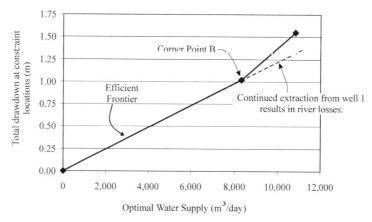

FIGURE 5.4   Total drawdown increases as water supply increases. Changes in slope correspond to corner point solutions.

5.2 RANGE ANALYSIS

Range analysis provides information about the local behavior of the optimal solution and its sensitivity to changes in parameter values. The analysis determines the range within which an objective function coefficient or a right-hand side value can vary without changing the optimal basis. These ranges are valid assuming all other parameters remain unchanged. Although the ranges are determined using the equations that follow, the process can be thought of as perturbing one parameter at a time about its original value until the basis changes.

Additional information obtained during range analysis consists of the variables that enter and leave the basis when the parameter of interest is perturbed beyond its range. The entering and leaving variables represent either a shift in resource utilization (if both are slack variables) or a change in the decision variables and associated active stress points. This information can be used to make inferences about solution robustness.

For the groundwater flow management problem, the constraint coefficients are a function of the simulation model and will not be treated with range analysis (see Section 6.6 for methods to incorporate simulation model parameter uncertainty). Nevertheless, there will be uncertainty in objective function coefficients and in the appropriate right-hand side values that will result in the desired aquifer response. This section presents the mathematics and discusses the interpretation of range analysis for objective function coefficients and right-hand side values. One of the benefits of using the simplex method is that the linear program does not have to be resolved several times to perform this analysis. Instead, the equations developed in Chapter 4 are used to determine the range analysis information. Recall that the derivations in Chapter 4 were conducted with the assumption that the decision variables had lower bounds of zero but no upper bounds. Because the equations in this section are based on the analysis from Chapter 4, these assumptions apply here as well. Upper bound constraints are implemented in MODOFC, which is on the CD-ROM included with this book, and the relevant simplex method and range analysis calculations are shown in the accompanying documentation.

5.2.1 OBJECTIVE FUNCTION COEFFICIENT RANGES

Range analysis on the objective function coefficients determines the range of coefficient values for which the optimal basis does not change. Recall from equation (4.25) that the optimal values of the decision variables depend only on the basis matrix. Thus, range analysis determines the coefficient ranges for which the pumping solution remains the same and the binding constraints do not change. At the same time, changes in the objective function coefficients will alter the value of the objective function and the shadow prices (Eq. 5.5).

Consider a change in the objective function coefficient for variable j. The new objective function is represented as

$$\text{maximize } f = \left(\underset{\sim}{c} + \Delta c_j e_j \right)^t x \tag{5.10}$$

where Δc_j is the coefficient change and e_j is a vector of zeros with a 1 in the jth row. In range analysis, the goal is to find the values of Δc_j for which the optimal solution, x^*, of the original problem remains unchanged.

Recall from Section 4.3.2 that the optimal solution is calculated from

$$x_B = B^{-1}(b - Nx_N) \tag{5.11}$$

and the objective function value at the optimal solution is determined as

$$f = c_B^t B^{-1} b + \left(\underset{\sim}{c}^t - c_B^t B^{-1} \underset{\sim}{A} \right) x \tag{5.12}$$

Note that the reduced costs $(\underset{\sim}{c}^t - c_B^t B^{-1} \underset{\sim}{A})$ are functions of the objective function coefficients and must be nonpositive at the optimum. This optimality condition provides the relationship for determining the coefficient ranges.

The mathematics for determining the coefficient ranges for nonbasic and basic variables differ, so we first derive the equations for a nonbasic variable j. Again, its reduced cost must remain nonpositive for the basis to stay unchanged, so the following relationship must hold:

$$-\infty \leq \left[(c_j + \Delta c_j) - (c_B^t B^{-1} \underset{\sim}{A}_j) \right] \leq 0 \tag{5.13}$$

where A_j is the jth column of the original coefficient matrix A. Rearranging (5.13) to solve for Δc_j,

$$-\infty \leq \Delta c_j \leq -\left(c_j - c_B^t B^{-1} A_j\right) \qquad (5.14)$$

Note that the upper bound is simply the negative of the reduced cost for nonbasic variable j. When the objective function coefficient reaches a value given by the bound in (5.14), the variable j will enter the basis. The leaving variable is determined using the ratio test described in Section 4.3.2.

In the water supply example, both candidate locations are active at the optimal solution, therefore they are both basic variables. However, if one of the decision variables were nonbasic, the foregoing equation would indicate the sensitivity of the optimal solution to the unit pumping cost assigned to the well. If Δc is small, then it might be important to reduce uncertainty in the coefficient. Conversely, if Δc is large, then the optimal solution would not change if the cost coefficient were to be slightly different than that used in the original problem. A similar analysis can be performed for the entering and leaving variables. If the entering or leaving variables are both slacks, then no stresses will be added to or removed from the basis. Conversely, if an entering or leaving variable is a stress, then the system of stress locations may not be robust to changes in the objective function coefficients.

The objective function coefficient ranges for basic variables require more computation because the reduced costs of all nonbasics, which must remain nonpositive, are functions of the objective function coefficients of the basic variables. Therefore, the reduced costs of all nonbasics must be checked during range computations for coefficients on each basic variable. Here, the index i will refer to the subset of variables that are basic. To determine the coefficient range for basic variable i, consider the reduced cost for nonbasic variable j

$$-\infty \leq \left[c_j - (c_B + \Delta c_i e_i)^t B^{-1} A_j\right] \leq 0 \qquad (5.15)$$

where e_i is an $(m \times 1)$ vector of zeros with a 1 in the ith row. Rearranging,

$$-\infty \leq -c_B^t B^{-1} A_j - \Delta c_i B_i^{-1} A_j \leq -c_j \qquad (5.16)$$

where B_i^{-1} is the ith row of B^{-1}. Further rearranging yields

$$\infty \geq \Delta c_i B_i^{-1} A_j \geq \left(c_j - c_B^t B^{-1} A_j\right) \qquad (5.17)$$

Recognizing that the element $(B_i^{-1} A_j)$ can be positive or negative and that the term on the right side is the nonpositive optimal reduced cost, the limits on Δc_i can be determined. Furthermore, (5.17) must be satisfied for all nonbasic variables, which dictates that we must find the limiting cases for determining Δc_i

$$\text{Max}_j\left\{\frac{(c_j - \mathbf{c}_B^t\mathbf{B}^{-1}\mathbf{A}_j)}{(\mathbf{B}_i^{-1}\mathbf{A}_j)} \,\middle|\, (\mathbf{B}_i^{-1}\mathbf{A}_j) > 0\right\} \leq \Delta c_i \qquad (5.18)$$

$$\leq \text{Min}_j\left\{\frac{(c_j - \mathbf{c}_B^t\mathbf{B}^{-1}\mathbf{A}_j)}{(\mathbf{B}_i^{-1}\mathbf{A}_j)} \,\middle|\, (\mathbf{B}_i^{-1}\mathbf{A}_j) < 0\right\}$$

The variable j that produces the limiting ratio for each bound is the nonbasic variable that will enter the basis. The leaving variable is again determined by the ratio test described in Section 4.3.2.

EXAMPLE OF COST RANGE ANALYSIS

For the water supply example, the maximum water supply is sought, so each candidate well has an objective function coefficient of one. Assume that this coefficient represents profit and that we want to maximize the profit obtained from supplying water. Range analysis can be done using the optimal solution determined in Section 4.4.2 and the values in equations (4.47), (4.48), and (4.49). Although the algebra for computing the ranges is not shown here, the range analysis for well 2 indicates that if the unit profit for that well fell to $0.70/(m^3/day)$, then it would no longer be beneficial to pump from that well. Instead, pumping would shift entirely to well 1 and constraint 1 would remain the only binding constraint. Therefore, the basic variables would be q_1, x_{s2}, and x_{s3} and the optimal solution would be at the corner point labeled B in Figure 4.1.

5.2.2 RIGHT-HAND SIDE RANGES

As in the previous section, the goal of right-hand side range analysis is to find the range of each right-hand side for which the basis does not change. For the objective function coefficient ranges, the reduced costs are used to determine when the optimal basis changes. The reduced costs are not functions of the right-hand side values, however, so a different approach is needed here. For this analysis, the nonnegativity constraints are invoked by recognizing that the optimal basis will not change as long as the basic variables remain nonnegative. Furthermore, because all basic variables are functions of the right-hand side values, a relationship can be defined for determining the ranges:

$$\{\mathbf{x}_B = \mathbf{B}^{-1}[(\mathbf{b} + \Delta b_k\mathbf{e}_k) - \mathbf{N}\mathbf{x}_N]\} \geq \mathbf{0} \qquad (5.19)$$

where $\mathbf{e}_k$ is an $(m \times 1)$ vector of zeros with a 1 in the kth row which represents the kth constraint.

Recall that the presentation assumes that decision variables all have lower bounds of zero, which results in setting the nonbasic variables to zero. Therefore, the term in (5.19) with the nonbasics can be eliminated. Consider the variable that is basic in row i

$$\{x_i = [(\mathbf{B}^{-1}\mathbf{b})_i + \Delta b_k \mathbf{B}_{ik}^{-1}]\} \geq 0 \tag{5.20}$$

This equation can be used to put bounds on the right-hand side of constraint k. Again, all basic variables must remain nonnegative for the basis to remain unchanged. Therefore, (5.20) must be applied to all constraints, with the limiting ratios used to identify the bounds on Δb_k:

$$\mathrm{Max}_i \left\{ \frac{-(\mathbf{B}^{-1}\mathbf{b})_i}{\mathbf{B}_{ik}^{-1}} \,\middle|\, (\mathbf{B}_{ik}^{-1}) > 0 \right\} \tag{5.21}$$

$$\leq \Delta b_k \leq \mathrm{Min}_i \left\{ \frac{-(\mathbf{B}^{-1}\mathbf{b})_i}{\mathbf{B}_{ik}^{-1}} \,\middle|\, (\mathbf{B}_{ik}^{-1}) < 0 \right\}$$

For both lower and upper bounds, the variable basic in row i that provides the limiting ratio will leave the basis.

To determine the entering basic variable, consider the constraint in row i corresponding to the leaving variable

$$x_{i,B} + (\mathbf{B}^{-1}\mathbf{N})_i^t \mathbf{x}_N = (\mathbf{B}^{-1}\mathbf{b})_i \tag{5.22}$$

and the current objective function, which contains the reduced costs

$$f(\mathbf{x}) = \mathbf{c}_B \mathbf{B}^{-1}\mathbf{b} + (\mathbf{c}^t - \mathbf{c}_B^t \mathbf{B}^{-1}\mathbf{\underset{\sim}{A}})\mathbf{x} \tag{5.23}$$

Recall that the reduced costs of basic variables are zero and nonbasic variables at the optimal solution have nonpositive reduced costs. If variable i is to leave the basis, then we must determine which variable j enters. A row operation is performed between the constraint corresponding to the leaving variable and the objective function in order to determine which reduced cost goes to zero first. This computation requires multiplying the constraint i by some value, γ, and subtracting this from the objective function. For all variables j, this computation must result in a nonpositive value in order to maintain optimality

$$\left(\mathbf{c}^t - \mathbf{c}_B^t \mathbf{B}^{-1}\mathbf{\underset{\sim}{A}}\right)_j - \gamma(\mathbf{B}^{-1}\mathbf{N})_{ij} \leq 0 \tag{5.24}$$

Rearranging (5.24) provides a limit on the multiplier γ

$$\gamma \geq \left\{ \frac{\left(\mathbf{c}^t - \mathbf{c}_B^t \mathbf{B}^{-1} \underset{\sim}{\mathbf{A}}\right)_j}{(\mathbf{B}^{-1}\mathbf{N})_{ij}} \,\middle|\, (\mathbf{B}^{-1}\mathbf{N})_{ij} > 0 \right\} \tag{5.25}$$

$$\gamma \leq \left\{ \frac{\left(\mathbf{c}^t - \mathbf{c}_B^t \mathbf{B}^{-1} \underset{\sim}{\mathbf{A}}\right)_j}{(\mathbf{B}^{-1}\mathbf{N})_{ij}} \,\middle|\, (\mathbf{B}^{-1}\mathbf{N})_{ij} < 0 \right\}$$

Because the reduced costs in the numerator are nonpositive, the first inequality indicates that the multiplier must be larger than zero. The second inequality must be tested for all nonbasic variables, with the limiting ratio indicating which reduced cost hits zero first. This is the entering basic variable. Letting the leaving variable identified above be basic in row i, the entering variable provides the limiting value in the following test:

$$\underset{j}{\mathrm{Min}} \left[\left(\frac{\left(\mathbf{c}^t - \mathbf{c}_B^t \mathbf{B}^{-1} \underset{\sim}{\mathbf{A}}\right)_j}{(\mathbf{B}^{-1}\mathbf{N})_{ij}} \right) \,\middle|\, (\mathbf{B}^{-1}\mathbf{N})_{ij} < 0 \right] \tag{5.26}$$

EXAMPLE OF RIGHT-HAND SIDE RANGE ANALYSIS

For the water supply example, range analysis can be calculated using the equations derived in this section and the appropriate values from Section 4.4.2. For constraint 1, the head bound can vary between 504.81 and 504.98 meters without changing the basis. At the lower head bound, constraint 2 becomes binding and the slack variable for constraint 3 enters the basis. At the upper head bound, well 1 stops pumping and slack variable 3 enters the basis. The head bounds can be substantiated from the data shown in Figure 5.3, which were determined during sensitivity analysis. Each change in slope in the figure implies a basis change. Range analysis finds the first basis change on either side of the original constraint value, which can be seen on the graph as slope changes at head values equal to 504.8 and 505.0 meters. Whereas the trade-off curve indicates when the basis will change, range analysis provides the additional information concerning which variables enter and leave the basis.

5.3 THE DUAL PROBLEM

For every linear programming problem, there is a corresponding dual problem that uses the same coefficients as the original problem. The solution to both formulations is the same, but the interpretation of the problems is different and additional insight is gained by investigating the meaning of both problems. For an original, or primal, problem of the form of our water supply example

$$\text{maximize} \quad f(\mathbf{q}) = \mathbf{c}'\mathbf{q} \tag{5.27}$$
$$\text{such that} \quad \mathbf{Aq} \geq \mathbf{b}$$
$$\mathbf{q} \geq \mathbf{0}$$

the dual takes the form

$$\text{minimize} \quad f(\boldsymbol{\lambda}) = \mathbf{b}'\boldsymbol{\lambda}$$
$$\text{such that} \quad \mathbf{A}'\boldsymbol{\lambda} \geq \mathbf{c} \tag{5.28}$$
$$\boldsymbol{\lambda} \leq \mathbf{0}$$

where $\boldsymbol{\lambda}$ represents the dual variables. The direction of the inequalities in the dual problem depends on the form of the primal, so not all dual problems will have constraints as shown here. The form of the dual program can be derived from the primal problem and the optimality conditions; this type of derivation for the linear program in (5.27) is shown in the next section.

5.3.1 DERIVING THE DUAL

The dual linear programming problem can be derived from the primal problem by using the optimality conditions developed in Section 4.3.2 and the definition of shadow prices given in Section 5.1.2. Shadow prices were defined as the local sensitivity of the objective function to the right-hand side values, or

$$\frac{\partial f}{\partial b_i} = \mathbf{c}_B^t \mathbf{B}_i^{-1} \tag{5.29}$$

Although the shadow prices are independent of the values of the decision variables, they are a function of the optimal basis. From this perspective, the linear programming solution process can be thought of as finding the optimal basis and thus the optimal shadow prices. In this context, the shadow prices can be viewed as the dual variables that must be optimized, along with the original, or primal, variables, during the solution process. By the definition in (5.29), there are m shadow prices, which can be represented in vector notation by the variable $\boldsymbol{\lambda}$ so that

$$\boldsymbol{\lambda} = \mathbf{c}_B^t \mathbf{B}^{-1} \tag{5.30}$$

In Section 4.3.2, the objective function at any iteration is shown to be

$$f(\mathbf{x}) = \mathbf{c}_B^t \mathbf{B}^{-1}\mathbf{b} + \left(\underset{\sim}{\mathbf{c}^t} - \mathbf{c}_B^t \mathbf{B}^{-1}\underset{\sim}{\mathbf{A}} \right) \mathbf{x} \tag{5.31}$$

Substituting the definition of the dual variables from (5.30) into (5.31)

$$f(\boldsymbol{\lambda}) = \boldsymbol{\lambda}^t \mathbf{b} + \left(\underset{\sim}{\mathbf{c}}^t - \boldsymbol{\lambda}^t \underset{\sim}{\mathbf{A}} \right) \mathbf{x} \tag{5.32}$$

The second term in (5.32) is always zero, and the objective function in terms of the dual variables is expressed as

$$f(\boldsymbol{\lambda}) = \mathbf{b}^t \boldsymbol{\lambda} \tag{5.33}$$

Recall that the optimality condition for the primal problem is that all reduced costs must be nonpositive. Constraints on the dual variables are imposed by the optimality condition for the primal. First, separate the reduced cost equation into the original decision variables and the surplus variables (recall that surplus variables are subtracted from the constraints)

$$\left(\underset{\sim}{\mathbf{c}}^t - \mathbf{c}_B^t \mathbf{B}^{-1} \underset{\sim}{\mathbf{A}} \right) \mathbf{x} = \left(\mathbf{c}^t - \boldsymbol{\lambda}^t \mathbf{A} \right) \mathbf{q} + \left(\mathbf{0}^t - \boldsymbol{\lambda}^t (-\mathbf{I}) \right) \mathbf{x}_s \tag{5.34}$$

Now apply the optimality condition

$$\left(\mathbf{c}^t - \boldsymbol{\lambda}^t \mathbf{A} \right) \leq \mathbf{0} \quad \text{and} \quad \left(\mathbf{0}^t - \boldsymbol{\lambda}^t (-\mathbf{I}) \right) \leq \mathbf{0} \tag{5.35}$$

and rearrange

$$\mathbf{A}^t \boldsymbol{\lambda} \geq \mathbf{c} \tag{5.36}$$

$$\boldsymbol{\lambda} \leq \mathbf{0}$$

Equations (5.36) provide a set of constraints on the values of the dual variables.

The remaining question is whether the dual objective in (5.33) should be minimized or maximized. To determine this, recall that at any nonoptimal but feasible primal corner point, at least one of the reduced costs is positive. This implies that the dual constraints (5.36) are not satisfied, further implying that the only feasible solution to the primal problem that is also feasible in the dual formulation is the optimal solution. Thus, the optimal solution of the primal problem is the minimum feasible solution in the dual problem, and the dual must be minimized.

5.3.2 INTERPRETING THE DUAL

To understand the dual formulation, we must take a closer look at the units of the variables and coefficients in the primal and dual problems. The primal and dual formulations for the water supply example are restated

$$\text{maximize} \quad f(\mathbf{q}) = \mathbf{c}^t \mathbf{q} \tag{5.37}$$

$$\text{such that} \quad \mathbf{A}\mathbf{q} \geq \mathbf{b}$$

$$\mathbf{q} \geq \mathbf{0}$$

$$\text{minimize} \quad f(\boldsymbol{\lambda}) = \mathbf{b}^t \boldsymbol{\lambda} \qquad (5.38)$$

$$\text{such that} \quad \mathbf{A}^t \boldsymbol{\lambda} \geq \mathbf{c}$$

$$\boldsymbol{\lambda} \leq \mathbf{0}$$

Although we have not presented the management linear program as a problem in managing limited resources, the dual is best interpreted from this perspective. In the context of the water supply example, the primal constraints place lower bounds on the head at specified locations in the model domain. Lower head bound constraints are equivalent to placing limits on the allowable drawdown. The bounds on drawdown can be interpreted as the maximum amount of the water resource that can be used by the pumping solution. The primal program can then be restated in economic terms as finding the maximum benefit from pumping without exceeding the resources available. From this perspective, the coefficient c_j is the benefit gained per unit stress from location j and q_j is the stress applied at well j. The constraint coefficients a_{ij} represent the change in head at location i per unit stress from location j and provide a measure of drawdown consumption. Each constraint states that the total drawdown at location i that results from stress from all locations j must not exceed the drawdown capacity at location i, given by b_i.

The shadow prices, or dual variables, have already been defined as the sensitivity of the original objective function to changes in the original constraints. Using the definitions in the previous paragraph, the dual variables can also be seen to represent the marginal value of drawdown at location i and have units of value or cost of drawdown consumption per unit drawdown at location i. The dual problem seeks to minimize the total cost of drawdown consumption across all locations i.

Returning to the primal problem, at any iteration, the reduced costs are

$$\text{reduced costs} = \left(\underset{\sim}{\mathbf{c}}^t - \boldsymbol{\lambda}^t \underset{\sim}{\mathbf{A}} \right) \qquad (5.39)$$

Using the dual interpretation, a solution to the primal problem is seen to be suboptimal if the marginal benefit of pumping from well j, c_j, is larger than the marginal cost of drawdown consumed by well j, $\boldsymbol{\lambda}^t \underset{\sim}{\mathbf{A}}_j$.

5.4 NOTES AND REFERENCES

Elements of the sensitivity analyses described in Section 5.1 are demonstrated by numerous authors including Peralta and Killian (1985), Ahlfeld *et al.* (1995), Reichard (1995), and Nishikawa (1998). Further discussion of the evaluation of slack variables, shadow prices, and reduced costs can be found in Bradley *et al.* (1977) and Hillier and Lieberman (1995).

Several studies have presented methods for finding the efficient frontier in multiobjective programming problems described in Section 5.1.3. A water supply problem complicated by saltwater intrusion concerns was analyzed by Shamir *et al.* (1984) in a multiobjective framework. El Magnouni and Treichel (1994) and Duckstein *et al.* (1994) solved problems in which water supply was maximized and costs and risk were minimized. Shafike *et al.* (1992) describe three techniques for selecting alternatives from a set of efficient solutions to a water supply problem that also sought to contain a contaminant plume at minimum cost. Ritzel *et al.* (1994) and Watkins and McKinney (1997) evaluate multiobjective formulations in plume containment problems.

The range analysis described in Section 5.2 is based on Bradley *et al.* (1977) and Nazareth (1987). Range analysis is implemented in the MODOFC code found on the CD-ROM included with this book. The range analysis equations for formulations with upper bounds on the decision variables may be found in the MODOFC documentation.

The dual program approach described in Section 5.3 was introduced by Gorelick (1982) and Gorelick and Remson (1982), who used it for a waste disposal problem. The dual was introduced as a means of improving the computational aspects of the problem as well as investigating potential disposal sites (i.e., the well location problem). Gorelick (1982) also presented results of a sensitivity analysis to investigate the effect of holding one injection rate constant on total disposal capacity. An interpretation of the dual problem associated with the management formulation for plume capture design and for water supply has been presented by Ahlfeld (1998).

6

ADVANCED LINEAR
FORMULATIONS

The linear groundwater flow management formulation presented in Chapter 3 is limited to the case of confined, steady-state groundwater flow, where constraints and objective functions are linear with respect to stress and hydraulic head. In this chapter, additional features that can be included in management formulations are introduced. These features follow the same general linear form as introduced in (4.1) and include accommodation of special features of the stress decision variables, formulations for transient groundwater flow and management, and incorporation of special constraint types. The general framework for constructing the groundwater flow management formulation remains unchanged. Some of the specific features of the solution process described in Chapters 4 and 5 may be affected and are described as they arise.

6.1 MANAGEMENT OF TRANSIENT FLOW

The formulations presented in previous chapters involve functions containing the steady-state hydraulic head generated by a steady stress. The introduction of a fixed duration stress or the potential for varying the stress rates over the life of the system requires reformulating the optimization problem. Simulating transient groundwater flow also has implications for the formulation. In this section, the implications of introducing a temporal component to the formulation are explored.

First we distinguish between three different discretizations of time. The smallest discretization is the numerical time step. As discussed in Chapter 2, this is selected to ensure numerical stability and provide adequate

FIGURE 6.1   Three different discretizations of time are defined for hydraulic control of transient flow.

accuracy for simulating the transient head response to changes in simulated stresses and boundary conditions. The next level of discretization is the interval at which external stresses are modified. These are stresses that are independent of the management strategy, such as seasonal variation in precipitation-based recharge. These external stresses are not part of the management model and will be considered part of the boundary conditions for the simulation model. It is possible for boundary conditions to change at every numerical time step. However, it is more common for changes to occur at a less frequent interval, such as monthly, with numerical time steps selected at much finer intervals. It is conventional to assume that the changes in boundary conditions occur at the beginning of the interval and continue at a fixed value during the entire interval. The third level of temporal discretization is the management period, which is the interval over which the stress strategy, which is represented by the decision variables, will be applied. Management periods may be coincident with the changes in boundary conditions but need not be. A typical relationship between these three time intervals is depicted in Figure 6.1.

Based on these definitions for different temporal discretizations, several possible combinations of stress and boundary condition changes are possible. The boundary conditions may vary over time while a single management period is used. Alternatively, the boundary conditions may remain fixed over time while the applied stress varies over multiple management periods. More generally, both the boundary conditions and applied stresses vary over time. Any variation of stresses or boundary conditions requires a transient flow simulation and appropriate selection of numerical time step.

6.1.1 TRANSIENT MANAGEMENT FORMULATION

When transient simulations or multiple management periods are to be included in the formulation, the general linear form of the optimization problem (4.1) must be revised. In general, constraints may be imposed on heads at any time. Practically, this implies observing heads at the end of numerical time steps, which need not correspond to the end of management periods or changes in boundary conditions. In this formulation, the head will be observed at T_h different periods indexed by τ. A total of T_m management periods are considered with the index t used to identify the management period. In addition, all coefficients are indexed according to the decision variable, state variable, or constraint that they modify. The transient management formulation is stated as

$$\text{minimize} \quad f = \beta + \sum_{t=1}^{T_m}\sum_{j=1}^{n}\alpha_{j,t}q_{j,t} + \sum_{\tau=1}^{T_h}\sum_{i=1}^{l}\gamma_{i,\tau}h_{i,\tau} \tag{6.1}$$

$$\text{such that} \quad \sum_{t=1}^{T_m}(\alpha_{1,t,k}^{q}q_{1,t} + \alpha_{2,t,k}^{q}q_{2,t} + \cdots$$

$$+ \alpha_{j,t,k}^{q}q_{j,t} + \cdots + \alpha_{n,t,k}^{q}q_{n,t} + \beta_{t,k}^{q})$$

$$+ \sum_{\tau=1}^{T_h}(\alpha_{1,\tau,k}^{h}h_{1,\tau} + \alpha_{2,\tau,k}^{h}h_{2,\tau} + \cdots + \alpha_{i,\tau,k}^{h}h_{i,\tau}$$

$$+ \cdots + \alpha_{l,\tau,k}^{h}h_{l,\tau} + \beta_{\tau,k}^{h}) \geq 0$$

$$\text{for } k = 1,\ldots,m$$

The constraints are a general linear function of all decision variables and all observed heads. In this form, the constraints may be used to represent such temporal quantities as cumulative extractions or temporal changes in head. If a single period is used and heads are observed only at the end of the management period, then this formulation reduces to the original general linear form. This formulation may also be used if it is desired to observe heads during a transient formulation with a single management period ($T_m = 1$, $T_h > 1$) or if multiple management periods are used with a single head observation ($T_m > 1$, $T_h = 1$).

The Taylor series expansion, which is used to relate heads and stress, is extended to include multiple management periods. The vectors of stresses, $\mathbf{q}$ and $\mathbf{q}_0$, are now understood to include stresses at each candidate location and management period with elements $q_{j,t}$ and $q_{j,t}^{0}$, respectively. Note that the derivatives required in the Taylor series account for the impact of stress on heads at previous management periods. Although the temporal summation extends over all management periods, the derivatives of head with respect to stress will be zero when the time represented by t is greater than the time represented by τ. This follows from the fact that head in the aquifer at a time τ is a function of current and past conditions, not of future stresses. The Taylor series expansion now becomes

$$h_{i,\tau}(\mathbf{q}) = h_{i,\tau}^{0}(\mathbf{q}_0) + \sum_{j=1}^{n}\sum_{t=1}^{T_m}\frac{\partial h_{i,\tau}}{\partial q_{j,t}}(\mathbf{q}_0)(q_{j,t} - q_{j,t}^{0}) + \cdots \tag{6.2}$$

As discussed in Chapter 2, when confined flow is assumed and linear boundary conditions are used, the response of head to stress is linear even when flow is transient and external stresses vary temporally. Hence, the first-order Taylor series approximation is exact and can be used, in the same manner as described in Section 4.1, to transform (6.1) into a linear program of the form

$$\text{minimize} \quad f = \sum_{j=1}^{n} \sum_{t=1}^{T_m} c_{j,t} q_{j,t} \tag{6.3}$$

$$\text{such that} \quad \sum_{j=1}^{n} \sum_{t=1}^{T_m} a_{kj,t} q_{j,t} \geq b_k, \quad k = 1, \ldots, m$$

Finally, response coefficients can be computed in a manner similar to that described in Section 4.2

$$\frac{\partial h_{i,\tau}}{\partial q_{j,t}} \approx \frac{h_{i,\tau}(\mathbf{q}_{\Delta j,t}) - h_{i,\tau}(\mathbf{q}_0)}{q_{\Delta j,t} - q_{j,t}^0} \tag{6.4}$$

6.1.2 STRESS AND WATER VOLUME

The primary decision variable, $q_{j,t}$, is the stress at specified locations and has units of volume per time. For problems with fixed duration it may be more important to consider optimizing the volume pumped in each management period. This quantity can be determined by multiplying the stress by the length of the management period, Δm_t. An objective function that takes this form would be

$$f = \sum_{t=1}^{T_m} \sum_{j=1}^{n} \alpha_{j,t} \Delta m_t q_{j,t} \tag{6.5}$$

where $\alpha_{j,t}$ now has dimension of monetary value per volume.

6.1.3 TIME VALUE OF MONEY

Another consideration when using multiple management periods is the time value of money. Stress costs incurred in later years can have a lower present value than stress costs incurred in earlier management periods. The present value cost of stress over T_m periods and with an interest rate i per management period can be written as

$$\text{minimize} \quad \sum_{j=1}^{n} \sum_{t=1}^{T_m} c_{j,t} \left[(1+i)^{-t} \right] q_{j,t} \tag{6.6}$$

where $c_{j,t}$ is the cost of stress per unit rate at location j during time period t, and the expression in the brackets is the present worth factor from elementary economic analysis. This form assumes that the management periods are all of the same duration and that the costs are incurred at the end of the management period. For the special case in which a single stress is sought over multiple time periods (6.6) can be rewritten as

$$\text{minimize} \quad \sum_{j=1}^{n} \left(\sum_{t=1}^{T_m} c_{j,t} \left[(1+i)^{-t} \right] \right) q_j \tag{6.7}$$

Note that all terms within the large parentheses are independent of stresses and can be replaced by a single coefficient to yield an objective function of the form

$$\text{minimize} \quad \sum_{j=1}^{n} C_j q_j \tag{6.8}$$

Finally, if the costs are the same for each location in each time period, then the coefficient, C_j, can be eliminated from the objective function.

6.2 UNDETERMINED DIRECTION OF STRESS

Under some circumstances it may be desirable to provide a candidate location for aquifer stress where the direction of that stress (extraction or recharge) is not specified and is to be determined by the solution procedure. In most simulation models the direction of stress is indicated by the sign on the value of stress. We have adopted the convention that positive stress is extraction from the aquifer and negative stress is recharge to the aquifer.

If the direction of stress is unspecified, then the variable that represents stress at the candidate location would be bounded above and below zero as

$$q_j^{ru} \leq q_j \leq q_j^{eu} \tag{6.9}$$

where q_j^{ru} is the upper bound on recharge (a negative number) and q_j^{eu} is the upper bound on extraction (a positive number). Consider the case in which the objective is to minimize the total rate of stress, regardless of direction, when a negative lower bound is present. If a portion of the fomulation is

$$\text{minimize} \quad \sum_{j=1}^{n} q_j \tag{6.10}$$

$$\text{such that} \quad q_j^{ru} \leq q_j \leq q_j^{eu}$$

the solution algorithm will tend to select the maximum recharge so as to minimize the objective function. This unintended result can be corrected by reformulating the objective to minimize the absolute value of the stress

$$\text{minimize} \quad \sum_{j=1}^{n} |q_j| \tag{6.11}$$

This form will tend to move the total stress toward zero, which is the intent of the formulation statement. Use of (6.11) introduces a new problem, namely that the objective function now is nonlinear and contains a discontinuous derivative when the stress equals zero. That is, the derivative on the positive side of zero has a different value than the derivative on the negative side of zero. Such an objective function requires specialized algorithms to solve.

An alternative is to reformulate the problem by replacing the single stress q_j by the sum of two new variables that represent the extraction and recharge components at location j. The elements of the formulation that relate to the stress appear as

$$\text{minimize} \quad \sum_{j=1}^{n}(q_j^e + q_j^r) \qquad (6.12)$$

$$\text{such that} \quad q_j = q_j^e - q_j^r$$

$$0 \le q_j^e \le q_j^{eu}$$

$$0 \le q_j^r \le -q_j^{ru}$$

where q_j^e is the extraction rate at location j and q_j^r is the recharge rate. Note that both of these variables are defined to be positive. Hence, the stress q_j is defined by the difference in the values of these two variables and will be negative if q_j^r is greater than q_j^e. Once the net stress is identified, all other constraints that involve the stress decision variable can be evaluated. By minimizing the sum of these two positive variables, the intent of minimizing total stress is achieved. Note also that it is not possible for the solution to have both extraction and recharge active simultaneously at a single location. If both stress components are nonzero, then q_j will have an absolute value smaller than that of either the extraction or recharge rate. It follows that to achieve a given value of q_j the sum of extraction and recharge rates is minimized when one of the two rates is zero.

Objective and constraint terms that involve hydraulic head are linear functions of stress regardless of the direction of that stress when the assumption of a confined aquifer is used. This can be confirmed by examining an analytical solution to the groundwater flow equation. From zero stress, an increase in extraction produces the same magnitude change in head (with opposite sign) as an equal increase in recharge. Hence, the derivative of head with respect to stress, q_j, has the same constant value at whatever value (positive or negative) of q_j it is evaluated. (Discontinuities in the response of head to stress can arise from other characteristics of the simulation model as discussed in Section 6.5.)

6.3 STRESS OVER MULTIPLE CELLS

Stresses are often applied through facilities that extend over multiple cells in the numerical grid. Such facilities include horizontal drains, long-screened vertical wells, and clusters of wells in a single well field. A common feature of these facilities is that water is withdrawn or recharged to the entire facility, resulting in a single flow distributed among multiple grid cells. The challenge is to relate the stress for the entire facility to the stresses at each individual grid cell encompassed by the facility. In this section, we explore some approaches to accommodating these types of facilities. In the discussion to follow, it is assumed that the location of the multiple-cell facility is known. Methods for solving the location problem for multiple-cell facilities are discussed in Chapter 7.

6.3.1 HORIZONTALLY EXTENSIVE STRESS

Horizontally extensive stress is applied by facilities such as drains, horizontal wells, infiltration galleries, recharge basins, or clusters of wells. Given a horizontal facility that extends over n_H grid cells, the single rate of stress, Q_H, which represents total stress at the horizontal facility, is defined by the constraint

$$q_1 + q_2 + q_3 + \cdots + q_{n_H} = Q_H \qquad (6.13)$$

In this case, the total stress becomes the decision variable for the problem and it can be used in additional constraints or in the objective function.

For facilities that consist of a single connected structure, such as a drain, additional conditions must be imposed on the stress at each individual cell contained within the facility. These conditions must ensure that individual cell stresses are distributed in such a way as to produce heads that mimic the impact of the facility on the aquifer. This implies imposing a requirement on the relationship between aquifer heads that are adjacent to the stress facility.

Consider, for example, facilities that consist of a perforated pipe. It may be reasonable to assume that hydraulic head within the pipe is nearly uniform at any point in time along the length of the pipe. This would follow from the observation that any head gradient, required to produce flow within the pipe, would be much smaller than the head gradient in the aquifer immediately adjacent to the pipe. If it is further assumed that the head in the pipe controls the head in the aquifer in the immediate vicinity of the pipe (implying either that there is no appreciable head loss from flow through the walls of the pipe or that the head loss is uniform), then it can be assumed that the head in the aquifer along the length of the pipe is uniform.

Using this reasoning, the formulation will require that heads in the aquifer be identical along the length of the pipe. These head constraints are expressed as

$$h_i = h_{i+1}, \qquad i = 1, \ldots, n_H - 1 \qquad (6.14)$$

Because it is difficult to achieve precise equality between two numerical values, it is necessary to replace each of the equality constraints in (6.14) with a pair of inequality constraints of the form

$$h_i - h_{i+1} \leq \delta \qquad (6.15)$$
$$h_{i+1} - h_i \leq \delta$$

where δ is a small value that represents a required level of numerical precision.

If the change in head along the length of the pipe is significant, then (6.14) can be replaced by

$$h_i = \alpha_i h_{i+1}, \qquad i = 1, \ldots, n_H - 1 \qquad (6.16)$$

where α_i represents the fraction of change in head during flow between points i and $i + 1$. From pipe flow theory, head loss will depend upon the flow rate in the pipe. Note that if this dependence is included then (6.16) becomes a nonlinear function of stress. A reasonable assumption may be to assign a specified value to α_i independent of the stress.

6.3.2 VERTICALLY EXTENSIVE STRESS

Vertically extensive stress arises most frequently from wells that are screened over substantial distances. For many numerical models, these distances may span multiple model layers or even multiple aquifer units. In a manner similar to that for horizontally extensive stress, a single rate of stress, Q_w, which represents total stress at the vertical facility over n_w grid cells, is defined by the constraint

$$q_1 + q_2 + q_3 + \cdots + q_{n_w} = Q_w \qquad (6.17)$$

The relationship between individual cell stresses can be constrained by the relationship between neighboring heads in a manner analogous to that for horizontally extensive facilities.

An alternative approach to defining the relationship between cell stresses is to assign proportions of flow based on the relative transmissivities of each numerical layer. The relationship that defines Q_w, (6.17), implies that the individual layer stresses can be defined as

$$q_j = \alpha_j Q_w, \qquad j = 1, \ldots, n_w \qquad (6.18)$$

where the α_j coefficients are bounded by zero and one and must satisfy $\sum_{j=1}^{n_w} \alpha_j = 1$. The coefficient α_j is defined so that flow from each layer is proportional to the transmissivity of that layer, or

$$\alpha_j = \frac{b_j K_j}{\sum_{j=1}^{n_w} b_j K_j} \tag{6.19}$$

where b_j is the thickness of the jth layer and K_j is the horizontal conductivity in the jth layer. This form assumes that the hydraulic gradient in the vicinity of the well is the same in each layer. This will be a reasonable assumption if the head in the well is nearly uniform, regardless of stress regime, and if the heads in the aquifer are also nearly uniform vertically.

6.4 CONSTRAINTS ON GROUNDWATER VELOCITY

Head difference constraints introduced in Section 3.2.2 are often used to produce flow in desired directions. With further examination, these constraints can be viewed as imposing requirements on the direction and magnitude of velocity at specified locations. Simply specifying that the head difference between two points is greater than some value guarantees only that a component of the velocity will be in the direction defined by a line between the two points.

Consider Figure 6.2a, where a positive head difference is required in the x-coordinate direction from point 1 to point 2. The velocity vectors indicated by the arrows in Figure 6.2a can all be characterized by their components in the x- and y-coordinate directions and the angle between the vector and the x-axis as depicted in Figure 6.2b. The velocity component in the x-direction, V_x, is positive for all vectors shown in Figure 6.2a, indicating that the vectors could satisfy the head difference constraint. In fact, any vector that falls within the half-plane defined as normal to the line connecting points 1 and 2 can satisfy the head difference constraints.

A single head difference constraint also fails to specify the magnitude of velocity. The relationship between a velocity component, V_x, and the angle θ between the component and resultant velocity vector, V, is given by $V = V_x / \cos \theta$. If a requirement is placed on the magnitude of V_x then, as θ increases from zero, the magnitude of velocity must also increase to satisfy those requirements. Without prior knowledge of the velocity direction, constraints on its magnitude are not possible when using a single head difference constraint. Only if the velocity direction is precisely specified can magnitude be directly bounded. Hence, a single head difference constraint imposes only a weak requirement on direction and an indirect requirement on the relationship between velocity magnitude and direction.

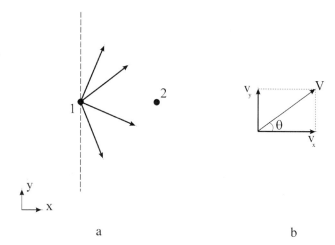

a   b

FIGURE 6.2 Requiring the head at 1 to be greater than the head at 2 contrains velocity to lie within the half-plane to the right of the dashed line in (a). A velocity vector in two dimensions can be defined by the identities shown in (b).

A more refined requirement on velocity can be determined by using multiple head difference constraints adjacent to each other. This approach relies upon the intersection of half-planes to further constrain the allowable direction of velocity. For example, two head difference constraints on the rectangular grid shown in Figure 6.3, one imposed between points 1 and 2 and the other imposed between points 1 and 3, can together produce requirements that will force the velocity vector to fall within the intersection

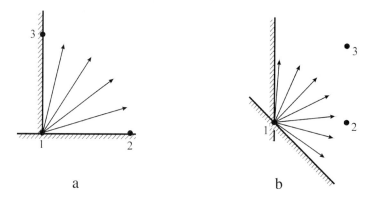

a   b

FIGURE 6.3 Multiple head difference pairs can be used to contrain velocity direction. Feasible directions in (a) lie within a 90 degree quandrant. Flow in (b) can be anywhere within the 135 degrees shown.

of the half-plane for each of the individual head difference constraints. For Figure 6.3a the direction of the velocity vector can be specified to within 90 degrees, in Figure 6.3b the direction can be specified to within 135 degrees.

The use of several constraints to control the direction of velocity can be generalized by defining two vectors A and B that each define half-planes. By specifying the direction of these two vectors, a range of allowable velocity directions can be defined. Figure 6.4 shows an arrangement of A and B that yields a particular range of velocity directions. The orientations of the vectors A and B are not limited to node points on the grid. Different orientations of A and B can provide arcs ranging from zero to 180 degrees within which the velocity may be constrained to lie.

The velocity constraints are defined by imposing requirements on the inner product between V, the velocity to be determined by the optimization, and the vectors A and B. From elementary linear algebra, if two vectors have a positive inner product then one lies in the half-plane defined by the other. Constraints are imposed to ensure positive inner products as follows:

$$V \cdot A = V_x A_x + V_y A_y \geq 0$$
$$V \cdot B = V_x B_x + V_y B_y \geq 0$$

$$(6.20)$$

A velocity vector V that satisfies the inner product constraints in (6.20) will have a direction that falls in the intersection of the half-planes defined by A and B.

The velocity, V, that satisfies (6.20) is also guaranteed to have a magnitude greater than or equal to zero. However, imposing a positive nonzero lower bound directly on the magnitude of velocity presents a challenge.

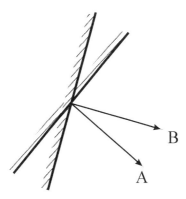

FIGURE 6.4 For any combination of vectors, A and B, the feasible velocity directions are defined by the intersection of half-planes.

Considering the relationship between the magnitude of V and its components

$$V = \sqrt{V_x^2 + V_y^2} \tag{6.21}$$

imposing a direct constraint on this magnitude would introduce a nonlinear function.

To retain linearity, bounds on the components of velocity can be imposed and would take the form

$$V_x \geq V_x^l; V_y \geq V_y^l \tag{6.22}$$

The presence of the bounds in (6.22) implies a bound on the magnitude of velocity of the form

$$V \geq \sqrt{(V_x^l)^2 + (V_y^l)^2} \tag{6.23}$$

Note that when both velocity components are at their lower bounds, the velocity direction will be fully defined.

If either of the velocity component bounds in (6.22) is not binding, then the direction of velocity may take a range of values; however, the presence of the bounds effectively reduces the set of feasible velocity directions and magnitudes. Figure 6.5 shows the region of feasible velocity directions and magnitudes when lower bounds on velocity are imposed. Here, the origin is the location at which velocity is to be constrained and the coordinate directions are the velocity components in the x- and y-directions. The intersection of the half-planes defined by the vectors A and B restricts the direction. Introducing lower bounds on the velocity components imposes

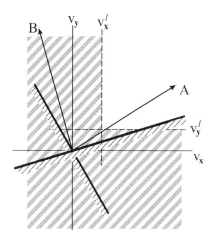

FIGURE 6.5   Lower bounds on velocity component magnitudes further restrict the region of feasible velocity vectors.

constraints on magnitude and further restricts feasible directions. The tip of the feasible velocity vector must lie in the unshaded region.

We now relate the velocity constraints to the general linear groundwater flow management form. The components of the velocity, V_x and V_y, depend upon the hydraulic heads, which in turn depend upon stress rates. This relationship can be most readily derived for a rectangular numerical grid that is aligned with the coordinate directions. Using the head difference notation from Section 3.2.2 and indicating three specified locations k_1, k_2, and k_3, the velocity components can be defined using head differences at specified node points

$$V_x = K\left(\frac{h_{k_1} - h_{k_2}}{\Delta x}\right)$$

$$V_y = K\left(\frac{h_{k_1} - h_{k_3}}{\Delta y}\right)$$

(6.24)

The equations in (6.24) indicate that the components of velocity are linear functions of hydraulic head. If it is assumed that hydraulic heads respond linearly to stress, then a first-order Taylor series can be used to relate velocity and stress as

$$V_x = V_x^0 + \sum_{j=1}^{n} \frac{\partial V_x}{\partial q_j} q_j$$

$$V_y = V_y^0 + \sum_{j=1}^{n} \frac{\partial V_y}{\partial q_j} q_j$$

(6.25)

where

$$\frac{\partial V_x}{\partial q_j} = \left(\frac{K}{\Delta x}\right) \frac{\partial \left(h_{k_1} - h_{k_2}\right)}{\partial q_j}$$

$$\frac{\partial V_y}{\partial q_j} = \left(\frac{K}{\Delta y}\right) \frac{\partial \left(h_{k_1} - h_{k_3}\right)}{\partial q_j}$$

(6.26)

Substituting (6.25) into (6.20) and rearranging yields

$$\sum_{j=1}^{n} \left(\frac{\partial V_x}{\partial q_j} A_x + \frac{\partial V_y}{\partial q_j} A_y\right) q_j \geq -V_x^0 A_x - V_y^0 A_y$$

$$\sum_{j=1}^{n} \left(\frac{\partial V_x}{\partial q_j} B_x + \frac{\partial V_y}{\partial q_j} B_y\right) q_j \geq -V_x^0 B_x - V_y^0 B_y$$

(6.27)

Finally, substituting (6.26) into (6.27) and rearranging results in

$$
\sum_{j=1}^{n}\left[\left(\frac{KA_x}{\Delta x}\right)\frac{\partial\left(h_{k_1}-h_{k_2}\right)}{\partial q_j}+\left(\frac{KA_y}{\Delta y}\right)\frac{\partial\left(h_{k_1}-h_{k_3}\right)}{\partial q_j}\right]q_j
$$

$$
\geq -V_x^0 A_x - V_y^0 A_y
$$

$$
\sum_{j=1}^{n}\left[\left(\frac{KB_x}{\Delta x}\right)\frac{\partial\left(h_{k_1}-h_{k_2}\right)}{\partial q_j}+\left(\frac{KB_y}{\Delta y}\right)\frac{\partial\left(h_{k_1}-h_{k_3}\right)}{\partial q_j}\right]q_j
$$

$$
\geq -V_x^0 B_x - V_y^0 B_y
$$

(6.28)

Examination of (6.28) indicates that the velocity constraints are a weighted combination of two individual head difference constraints. They take the same general linear form as (3.7) and so can be resolved into linear inequalities in the stress decision variables.

6.5 NONLINEAR BOUNDARY CONDITIONS

The formulations and solution procedures developed in Chapters 3 through 5 presume a linear response of head to stress. As discussed in Chapter 2, linear response is guaranteed when the aquifer is modeled as confined with linear boundary conditions. Many simulation models include the capability to couple the behavior of the groundwater flow system with other hydrologic features that may be connected to the aquifer. Such features may include surface water bodies, such as rivers or lakes, or head-dependent interactions with the unsaturated zone and the atmosphere. These interactions are typically introduced through nonlinear boundary conditions imposed on the aquifer domain. These boundary conditions may be nonlinear because of their quasi-linear form and may have discontinuous derivatives. In this section, we explore problems that can arise when these nonlinear boundary conditions are present.

Under some circumstances, these boundary conditions may cause a nonlinear response between head and stress, even when the aquifer is modeled as confined. For example, consider the interaction between an aquifer and a stream. If the aquifer and stream are hydraulically connected, then a head-dependent flux boundary condition is imposed. In this instance, the flux, q_{stream}, between the aquifer and stream is proportional, through a constant, C, to the difference in aquifer head adjacent to the stream, h_i, and stream head, h_{stream}. This relationship is expressed as

$$
q_{\text{stream}} = C\left(h_i - h_{\text{stream}}\right)
$$

(6.29)

However, when the aquifer and stream are not connected by a continuous saturated zone, a specified flux is used to represent the leakage from the stream to the aquifer. Such hydraulic disconnection might occur as a result of pumping from the aquifer. As the aquifer head decreases, a point is reached at which the saturated hydraulic connection is broken. This might be modeled as occurring when the aquifer head drops below the bottom of the streambed, b_{stream}. The resulting relationship between aquifer head and flux between the aquifer and stream is described by

$$q_{\text{stream}} = C(h_i - h_{\text{stream}}) \qquad \text{if } h_i \geq b_{\text{stream}}$$
$$q_{\text{stream}} = C_0(b_{\text{stream}} - h_{\text{stream}}) \qquad \text{if } h_i < b_{\text{stream}} \tag{6.30}$$

and depicted in Figure 6.6, where positive values of q_{stream} indicate water leaving the aquifer.

The nonlinear boundary condition depicted in Figure 6.6 is made up of two linear boundary conditions and includes a discontinuity in the derivative of stream flux with respect to aquifer head. The boundary condition is nonlinear because the stream flux regime and the resulting form of the boundary condition depend on aquifer head, which in turn depends on the stresses.

The solution of any optimization formulation is complicated if the simulation model that is coupled to the optimization contains nonlinear boundary conditions. If boundary conditions are piecewise linear and flow behavior is such that a single linear segment will always be the active boundary condition (e.g., if the stream will be hydraulically connected under all stress scenarios that the optimizer might test), then the nonlinear boundary

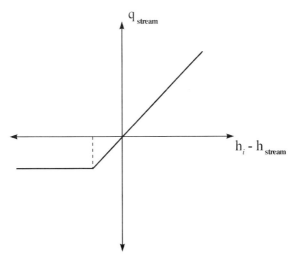

FIGURE 6.6 Example of a nonlinear boundary condition. When aquifer head, h_i, falls below a certain level, the aquifer and stream are no longer hydraulically connected and the boundary condition changes abruptly.

condition will behave like a linear boundary condition and linear solution methods can be used. In some cases, it may also be valid to assume that the boundary conditions have an insignificant impact on the relationship between head and stress. This might arise if the nonlinearity is mild or if the boundary condition is applied at a significant distance from the candidate locations and head observation points. However, if these conditions on the boundary conditions are not met, then other approaches are needed.

For a single location and a single head observation point it is possible to determine the value of stress at which the transition of boundary condition regimes will occur, suggesting that an algorithm could be constructed that would anticipate the change in boundary response. In the presence of multiple stress locations, however, aquifer head is affected by stress at all locations, making it impractical to identify all the possible combinations of stresses at which the transition will occur. Further, if the stream is represented by a series of cells at which aquifer head drives stream flux, then the change in head at any one point with respect to stress at any other point depends on the boundary condition regime at all stream cells. This response function would take the form of a piecewise linear function where each segment corresponds to a particular combination of boundary condition regimes at the various stream cells. As with the simpler case, the value of aquifer head at which the boundary condition makes the transition from one linear segment to the next depends on stress at all candidate locations.

6.6 ACCOUNTING FOR PARAMETER UNCERTAINTY

In Chapter 5, various methods for investigating the impact of uncertainty in the objective function coefficients and right-hand side values of linear management problems are presented. In this section, we introduce methods for accounting for uncertainty in simulation model parameters.

The optimization approach is based on determining the strategy that pushes the physical capabilities of the aquifer to their limits. Because of this, an optimal solution will nearly always be binding at one or more constraints that reflect the physical limits of the simulated aquifer. It follows that any error in the simulated aquifer behavior will produce a nonoptimal, and perhaps infeasible, solution when the strategy is implemented in the field. A strategy that is suboptimal relative to the management model is superior to a strategy that will prove infeasible when applied in the field.

A straightforward approach to accounting for uncertainty in simulation model parameters is the use of a safety factor on constraints. For example, if the design criteria require that head be below a specified level, then the constraint imposed may stipulate a head level below that actually required. This approach is meant to ensure that the intended constraint is satisfied when the strategy is applied in the field. A challenge with this approach is determining the appropriate value for the safety factor.

A useful technique for introducing a safety factor is the chance-constrained method, which scales the safety factor in relation to the uncertainty in aquifer parameters. In many settings the most significant uncertain parameter is the hydraulic conductivity. A common approach to characterizing uncertainty in the value of hydraulic conductivity is to represent conductivity as a stationary, correlated random field. This implies that conductivity is a random variable that is represented by a probability distribution.

With a characterization of the uncertainty in conductivity, a new formulation can be constructed using the method of chance constraints. The chance-constrained formulation presumes that the uncertainty lies in evaluating the constraints and requires that the constraints be satisfied with a specified probability. As an example, we write the compact form of the groundwater flow management formulation, where each constraint is represented by the symbol Λ_k

$$\text{minimize} \quad f = \sum_{j=1}^{n} \alpha_j q_j \tag{6.31}$$

$$\text{such that} \quad \sum_{j=1}^{n} a_{kj} q_j = \Lambda_k \geq b_k, \qquad k = 1, \ldots, m$$

For this case, presume that the objective function coefficients, α_j, are known with certainty but that the constraint coefficients are uncertain because they depend on hydraulic heads, which in turn depend on the uncertain hydraulic conductivity. It follows that the constraint, Λ_k, can be considered a random variable.

The chance-constrained formulation is written to include a bound on the probability that the constraint is satisfied:

$$\text{minimize} \quad f = \sum_{j=1}^{n} \alpha_j q_j \tag{6.32}$$

$$\text{such that} \quad \Pr[\Lambda_k \geq b_k] \geq p_k, \quad k = 1, \ldots, m$$

where $\Pr[\ ]$ implies probability and p_k is a specified probability level or reliability measure.

With appropriate assumptions about the characteristics of the constraints, the probabilistic formulation in (6.32) can be converted into a deterministic formulation with the reliability level as a user-defined parameter. One approach is to assume that the constraints Λ_k are normally distributed with mean μ_k and standard deviation σ_k and that the distribution of each constraint is independent of the stress. The probabilistic constraints can be transformed to a standard normal distribution by

$$\Gamma_1\left[\frac{\Lambda_k - \mu_k}{\sigma_k} \geq \frac{b_k - \mu_k}{\sigma_k}\right] \geq p_k \tag{6.33}$$

Under this transformation, each constraint becomes a standard normal random variable with zero mean and unit variance. The probability statement can now be replaced by a statement involving the standard normal cumulative probability function, F,

$$1 - F\left[\frac{b_k - \mu_k}{\sigma_k}\right] \geq p_k \tag{6.34}$$

Equation (6.34) can be rearranged using the inverse of the cumulative probability function to

$$F^{-1}[1 - p_k] \geq \frac{b_k - \mu_k}{\sigma_k} \tag{6.35}$$

Using the identity $F^{-1}[1 - p_k] = -F^{-1}[p_k]$ and rearranging yields the constraint

$$\mu_k \geq b_k + \sigma_k F^{-1}[p_k] \tag{6.36}$$

One approach to determining the parameters of the constraint distributions, μ_k and σ_k, is to employ a first-order Taylor series expansion of the constraint with respect to the random hydraulic conductivity such that

$$\Lambda(\mathbf{K}) = \Lambda(\mathbf{K}_0) + J_\Lambda(\mathbf{K} - \mathbf{K}_0) \tag{6.37}$$

where $\mathbf{K}_0$ is the vector of expected values of hydraulic conductivity and J_Λ is the Jacobian matrix of derivatives of each constraint function with respect to each of the elements of the hydraulic conductivity vector. Although not explicitly shown in (6.37), the Jacobian matrix is a function of the stress rates. Based on (6.37), we can derive

$$\mu_k = E[\Lambda_k] = \Lambda_k(\mathbf{K}_0) \tag{6.38}$$

and

$$\text{Cov}[\Lambda] = J_\Lambda \text{Cov}[\mathbf{K}]J_\Lambda \tag{6.39}$$

where $E[\Lambda_k]$ is the expected value of constraint k and $\text{Cov}[\Lambda]$ is the constraint covariance matrix. Equation (6.38) states that the expected value of the constraints is simply the constraints evaluated at the expected conductivity. Assuming that $\mathbf{K}_0$ is used in deterministic evaluation of the constraints, then μ_k is the deterministic value of the constraints. The standard deviation of the constraints, σ_k, can be extracted from (6.39).

The full chance-constrained formulation can now be stated by combining (6.36) and (6.38)

$$\text{minimize} \quad f = \sum_{j=1}^{n} \alpha_j q_j \tag{6.40}$$

$$\Lambda_k(\mathbf{K}_0) \geq b_k + \sigma_k F^{-1}[p_k]$$

The chance constraint (6.36) can now be interpreted as simply adding a safety factor, $\sigma_k F^{-1}[p_k]$, to the deterministic constraint. The standard deviation, σ_k, is a nonlinear function of the stress rates, so to make the constraints linear, we must assume that σ_k is constant. Without this assumption, it would be impossible to determine the statistical parameters without knowing the stress rates.

Although the chance constraint, as derived here, provides a useful interpretation of the safety factor, it does depend on several assumptions that may not be valid. The first assumption is that the constraints are normally distributed. This assumption implies that if the constraint were computed for a particular set of stresses under many possible parameter values, then the set of constraint values would be normally distributed. This assumption also implies that the standard deviation of the constraint distribution is identical for different sets of stresses.

The second assumption that is required in the preceding derivation is that the first-order approximation of the constraints (6.37) is accurate. It is generally believed that the accuracy of the first-order approximation decreases as the coefficient of variation of hydraulic conductivity increases. Inadequate research has been conducted to confirm the validity of either of these assumptions under a range of conditions, although it is likely that their validity will depend on both the source of the uncertainty and the nature of the constraints.

When using any form of safety factor, regardless of the means of determining its value, a further concern is the potential for introducing conflicting constraints through overconstraining. As discussed in Chapter 3, the density of constraints and their tendency to induce overconstraining must be considered. Judicious consideration of the interaction between constraint location and modified constraint value is recommended.

6.7 NOTES AND REFERENCES

The transient formulation described in Section 6.1 is a generalization of the formulation of Aguado and Remson (1974), who used a one-dimensional simulation model. Alley *et al.* (1976) included transient conditions in a two-dimensional model. Atwood and Gorelick (1985) extended the plume control formulation to include multiple management periods in which stresses and well locations could change at each period.

The method for linearizing an absolute value function described in Section 6.2 is a widely used technique drawn from the operations research literature (Bradley *et al.*, 1977).

Velocity constraints related to the form described in Section 6.4 have been introduced by several authors. Colarullo *et al.* (1984, 1985) and Heidari *et al.* (1987) amended the groundwater management formulation to include velocity constraints and applied the formulation to plume control. Contaminant plume cleanup times were considered by Lefkoff and Gorelick (1985) by including groundwater velocity constraints. Ahlfeld and Sawyer (1990) proposed a formulation in which the direction of the velocity at specified locations is constrained to lie at a specified angle by constraining the ratio of the velocity components. Ratzlaff *et al.* (1992) derived this result from a different perspective using specified angles to bound the direction of velocity. Their approach yields constraints similar to (6.27) but with the components on the direction-defining vectors A and B replaced by appropriate trigonometric functions of the specified angles.

The application of chance constraints to groundwater management described in Section 6.6 is drawn from Tiedeman and Gorelick (1993) and Wagner and Gorelick (1987). Chance constraints were first introduced into the operations research literature by Charnes and Cooper (1959). Incorporation of uncertainty into groundwater management models remains an active area of research. Approaches in which multiple realizations of the constraints must be simultaneously satisfied have garnered significant attention (Chan, 1993, 1994; Morgan *et al.*, 1993). Analyses of the impacts of uncertainty have also been conducted (Zhen and Uber, 1996; Verdon, 1995). Other general approaches are described by Tung (1986), Wagner and Gorelick (1989), Wagner *et al.* (1992), Watkins and McKinney (1997), and Sawyer and Lin (1998).

7

FORMULATIONS WITH BINARY VARIABLES

For many applications it becomes desirable to impose requirements on the solution that cannot readily be formulated using continuous variables. These requirements might include a stipulation that the total number of active stress locations be bounded within a specified range or that the stress at each candidate location be either zero or greater than a nonzero lower bound. It may also be desirable to incorporate discontinuous measures into the objective function, such as the cost of facility construction.

To incorporate elements of this type into the optimization formulation a new type of decision variable must be introduced: the integer variable. Integer variables are decision variables that are constrained to have integer values. The use of integer variables will be limited here to binary integer variables, where the variable may take only the value zero or one.

7.1 MIXED BINARY LINEAR FORMULATION

The linear formulation introduced in Section 3.2.4 is extended to include n_b binary variables, X_b. Binary variables will be used in the following examples to indicate the stress activity. The binary variables appear in the objective function, weighted by coefficients κ_b, and in the constraints. The mixed binary linear formulation is

$$\text{minimize} \quad f = \beta + \sum_{j=1}^{n} \alpha_j q_j + \sum_{i=1}^{l} \gamma_i h_i + \sum_{b=1}^{n_b} \kappa_b X_b \quad (7.1)$$

such that

$$
\alpha_{1,k}^q q_1 + \alpha_{2,k}^q q_2 + \cdots + \alpha_{j,k}^q q_j + \cdots + \alpha_{n,k}^q q_n + \beta_k^q
$$
$$
+ \alpha_{1,k}^X X_1 + \alpha_{2,k}^X X_2 + \cdots + \alpha_{b,k}^X X_b + \cdots + \alpha_{n_b,k}^X X_{n_b} + \beta_k^X
$$
$$
+ \alpha_{1,k}^h h_1 + \alpha_{2,k}^h h_2 + \cdots + \alpha_{i,k}^h h_i + \cdots + \alpha_{l,k}^h h_l + \beta_k^h \geq 0,
$$

$$\text{for } k = 1, \ldots, m$$

$$0 \leq X_b \leq 1 \qquad b = 1, \ldots, n_b$$

$$X_b \text{ is integer} \qquad b = 1, \ldots, n_b$$

Note that two additional constraints are added that bind the value of the binary variables to lie between zero and one and require that the binary variables be integer. This last constraint is referred to as the integrality constraint.

In the next section, a means of relating the binary variables to the stress variables is introduced and a number of possible new formulation elements that make use of binary variables are introduced. This is followed by a brief description of algorithms used to solve the mixed binary linear problem.

7.2 EXAMPLES OF MIXED
BINARY FORMULATIONS

In all of the examples given in this section, binary variables are used to indicate the presence or absence of stress at a specified location or facility or to indicate a range of stress at a single location. In every case, it is convenient to define the binary variable so that X_b is equal to one when the corresponding stress is nonzero and equal to zero if the stress is zero. Defined in this fashion, the binary variable can be used to indicate the need for a facility, such as a well, at which to apply the stress. If the stress at a specified location is nonzero, then the unit value of the associated binary variable indicates that a facility must exist. If the stress at a location is zero, then no facility is needed.

The key to using binary variables is to relate their value to the value of the stress. For the case in which the binary variable represents the presence of an extraction stress at a location, this relationship can be defined using constraints of the form

$$q_j \leq MX_j \tag{7.2}$$

where M is a large positive value. The subscript on the binary variable is selected to be the same as that on the corresponding stress. With this constraint imposed, X_j will be forced to a value of one if q_j is greater than zero. Standing alone, this constraint does not ensure that X_j will be zero when q_j is zero. Care must be taken that other formulation characteristics will tend to force X_j to zero when q_j is zero. Several examples of ways to accomplish this are described subsequently.

The integer definition constraint (7.2) will be effective if M is assigned any positive value. Note, however, that when X_j is equal to one this constraint implicitly serves as an upper bound on the value of q_j. For this reason, a useful definition of M is to assign it a value equal to the upper bound on stress at location j, although any value larger than the upper bound is acceptable.

7.2.1 INCLUDING FIXED CHARGE COSTS

The expanded objective function in (7.1) includes a new term involving the weighted sum of binary variables. This term can be used to represent fixed charges associated with the presence of each facility, where κ_b represents the fixed charge associated with facility b. The construction cost for the facility is an example of a fixed charge that might be included. It will be important to include construction costs in the formulation when they are significant relative to long-term operating costs.

Consider the objective function for a problem with a single stress location, where costs depend only on the stress rate and the fixed charge. This function would take the form

$$f = \alpha q + \kappa X \tag{7.3}$$

and is depicted in Figure 7.1. When q is zero then X is zero and both terms of the cost function are zero. When q takes on any nonzero value then X becomes one (the facility must be constructed) and the objective function jumps to a new value with costs increasing linearly as stress continues to increase. The resulting objective function is discontinuous. When κ is set to zero then the objective function returns to the continuous linear form described in Chapter 3.

It is important to note that when the binary variables are included in the objective function with positive-valued coefficients κ, the optimization formulation will tend to force the binary variables to zero values when the objective is to minimize the function. An objective function of this form, in combination with constraints of the form of (7.2), will guarantee that the binary variable will be nonzero if and only if the corresponding stress is nonzero.

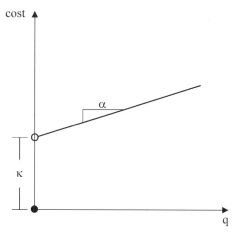

FIGURE 7.1   Cost function with installation cost, κ, and cost per unit pumping rate, α. In this case, the cost function is discontinuous at the origin.

7.2.2 NONZERO LOWER BOUNDS ON STRESS

As described in Section 3.2.1, upper and lower bounds can be placed on stresses at individual locations. In formulations used to solve the stress location problem it is required that the algorithm be provided the opportunity to set stresses at candidate locations to a zero value indicating that the corresponding location has not been selected. One approach to accommodating the stress location problem is to set the lower bound to zero. As described in Section 3.3.1, if a nonzero lower bound is used then, in effect, the formulation forces all candidate stresses to operate at a nominal level. Using a zero lower bound, however, allows the optimization algorithm to select stresses that may be too small to be practical.

The use of binary indicator variables provides an alternative approach to accommodate both meaningful lower bounds and the stress location problem. Using binary variables, it is possible to formulate constraints that are equivalent to requiring that the stress must be either zero or greater than some nonzero lower bound q_j^l. The constraints are effectively

$$q_j \geq q_j^l > 0 \tag{7.4a}$$

or

$$q_j = 0 \tag{7.4b}$$

The condition implied by (7.4) can be expressed using (7.2) and the additional constraint

$$q_j \geq q_j^l X_j \tag{7.5}$$

The binary constraint (7.2) forces X_j to be one when q_j is nonzero. When X_j is equal to one, then (7.5) takes the form of a nonzero lower bound constraint. When q_j is equal to zero then the corresponding binary variable X_j is also zero and (7.5) takes the form of a zero lower bound constraint.

7.2.3 COUNTING THE NUMBER OF ACTIVE LOCATIONS

Once the binary variables are defined, the sum of the binary variables is a direct measure of the number of locations where stress is nonzero, that is,

$$\sum_{b=1}^{n_b} X_b = \text{number of active stress locations} \qquad (7.6)$$

This measure can be used to form any number of constraints. Two simple examples are upper and lower bounds on the number of active stress locations. These constraints take the form

$$\sum_{b=1}^{n_b} X_b \leq N^u$$
$$\sum_{b=1}^{n_b} X_b \geq N^l \qquad (7.7)$$

where N^l and N^u are the respective bounds on the number of active locations.

7.2.4 LOCATION PROBLEM FOR
MULTIPLE-CELL FACILITIES

Multiple-cell facilities include horizontal wells, drains, and vertical wells that are screened over multiple numerical layers. Multiple-cell facilities are described in Section 6.3, where it is presumed that the location, orientation, and length of each facility are defined and that the only variable to be designed is the total stress for the facility (and implicitly the stresses at the individual cells that make up the facility). Binary variables can be incorporated into formulations for multiple-cell facilities in several ways discussed in the following to assist in designing the location, orientation, and length of these facilities.

A simple approach to this problem is to mimic the location methodology for single cell locations, namely to provide a number of candidate multiple-cell facilities and allow the optimization algorithm to select from among these. Consider a series of n_f candidate facilities where the location of the starting cell, orientation, and length are predetermined. Each facility is assigned an indicator variable, X_f. We first must ensure that the indicator variable is active when the facility is active. This can be accomplished by constraints of the form

$$Q_f \leq MX_f \qquad f = 1, \ldots, n_f \qquad (7.8)$$

where Q_f is the total stress for the facility as defined by either (6.13) or (6.17) in Section 6.3. If the candidate facilities overlap, then the indicator variable must be tied to the stress at nonoverlapping cell locations. As with the binary variable for a single cell stress, this constraint needs to be combined with other formulation characteristics that tend to drive X_f to zero, such as an objective that minimizes the cost of facility construction. Constraints similar to (7.7) may be added to limit the total number of facilities. It may also be desirable to provide candidate facilities, subsets of which are mutually exclusive. For example, two candidate locations may overlap with one longer than the other, or two locations may intersect. Under these circumstances it may be desirable to require that at most one of the overlapping facilities be selected. This can be ensured by imposing constraints of the form

$$\sum_{f \in f_i} X_f \leq 1 \qquad (7.9)$$

where f_i is the set of candidate facilities that are overlapping.

When binary variables are used to determine the location for multiple-cell stress facilities, steps must be taken to ensure that head constraints associated with specific facilities are imposed only if the facility is chosen. This can be accomplished by modifying constraints to be satisfied automatically if the corresponding facility is not active.

Consider a case with two overlapping candidate stress facilities. Assume that both facilities span four grid cells, three of which are common to both facilities as shown in Figure 7.2. There are three possible outcomes regarding these two facilities: 1) neither one of them is chosen, 2) facility 1 is chosen, or 3) facility 2 is chosen. The optimization formulation must account for these different possibilities. Restricting the discussion to just cells 1 and 2 shown in Figure 7.2, if facility 2 is chosen, the following head constraints must be satisfied:

$$\begin{aligned} h_1 - h_2 &\leq \delta \\ h_2 - h_1 &\leq \delta \end{aligned} \qquad (7.10)$$

However, if facility 1 is chosen or neither facility is chosen, these constraints must be transformed so that they are always satisfied and therefore do not influence the solution. To make this transformation, we can require that the two head differences be smaller than some very large value if facility 2 is not chosen. If we set this value sufficiently large, then the constraint will always be met and we will have, in effect, eliminated the constraint. To do this, we alter (7.10) as follows:

$$h_1 - h_2 \leq \delta + M(1 - X_2) \tag{7.11}$$

$$h_2 - h_1 \leq \delta + M(1 - X_2) \tag{7.12}$$

where M is a large value. If neither facility is chosen or if facility 1 is chosen, then $X_2 = 0$ and the second term on the right side of each constraint is set to M. Conversely, the second term on the right side of each constraint equals zero if facility 2 is chosen and the constraints become equivalent to (7.10).

A similar approach can be used to handle the head constraints across cells that are common to the overlapping but mutually exclusive facilities. Limiting the discussion to grid cells 2 and 3 in Figure 7.2, the head constraints

$$h_2 - h_3 \leq \delta$$
$$h_3 - h_2 \leq \delta \tag{7.13}$$

must be imposed if either facility is chosen. If neither facility is chosen, these constraints must be trivially satisfied. In this case, the constraints are altered with one additional term

$$h_2 - h_3 \leq \delta + M(1 - X_1 - X_2) \tag{7.14}$$

$$h_3 - h_2 \leq \delta + M(1 - X_1 - X_2) \tag{7.15}$$

As before, these constraints are satisfied if neither stress facility is active.

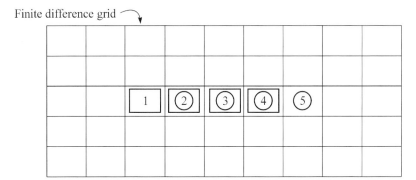

FIGURE 7.2 Example of multiple-cell horizontal stress facilities in plan view. Facility 1 is represented by the numerical grid cells with circles and facility 2 is represented by the cells with rectangles. The facilities are mutually exclusive; at most one of the two can be chosen for the optimal design.

7.2.5 MULTIPLE SEGMENT OBJECTIVE FUNCTIONS

Binary variables can be used for more complex situations, such as piecewise linear objective functions. Consider a situation in which the fixed charge and unit operational cost depends on the stress rate to be applied. Let a well and associated pump be installed at price κ_1 when operated at a stress that ranges from zero to an upper bound of q_1^u. If the stress must be greater than this upper bound, then a more expensive well and/or pump must be installed at a cost of κ_2. This larger capacity pumping facility can be operated between the range q_1^u and q_2^u.

The combined construction and operational cost as a function of stress at this location is depicted in Figure 7.3. Costs in the stress range zero to q_1^u have the same behavior as described in Figure 7.1, where α_1 is the cost per unit stress. If stress is required above a rate of q_1^u, then operational cost per unit stress is assigned a value α_2 that may differ from α_1. The unit cost required to increase stress above q_1^u consists of both the construction cost κ_2 and the operational costs at the stress q_1^u.

This problem can be placed in the general mixed integer linear form by defining two variables q_1 and q_2 to represent the stress components in each of the two ranges depicted in Figure 7.4. These variables will be defined so that the sum of the two is equal to the total stress

$$q = q_1 + q_2 \tag{7.16}$$

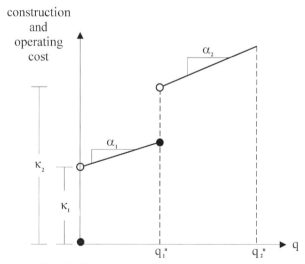

FIGURE 7.3   A piecewise linear cost function results when construction costs are a function of pumping rates.

Note that in Figure 7.4 a cost increment, $\widehat{\kappa}_2$, is defined that accounts for the increase in cost to move from segment 1 to segment 2. This cost function can now be stated as

$$f = \alpha_1 q_1 + \kappa_1 X_1 + \alpha_2 q_2 + \widehat{\kappa}_2 X_2 \tag{7.17}$$

The binary variables, X_1 and X_2, represent the presence or absence of stress in each of the two segments. They are each defined using their respective upper bounds

$$q_1 \leq q_1^u X_1 \tag{7.18}$$

and

$$q_2 \leq q_2^u X_2 \tag{7.19}$$

For this formulation to behave as we would like, the constraints must require that q_2 be nonzero only when q_1 is at its upper bound. Consider several situations. If $\alpha_1 \leq \alpha_2$ then minimization of (7.17) will tend to prefer activation of q_1 over q_2 until q_1 reaches its upper bound. However, if $\alpha_1 \geq \alpha_2$ then this formulation contains nothing to force q_1 to its upper bound before q_2 is activated. This problem can be resolved by adding an additional constraint to force the desired behavior explicitly

$$q_1 \geq q_1^u X_2 \tag{7.20}$$

If q_2 is nonzero, then (7.19) ensures that X_2 will be one. If X_2 is one then (7.20) ensures that q_1 must be greater than or equal to its upper bound.

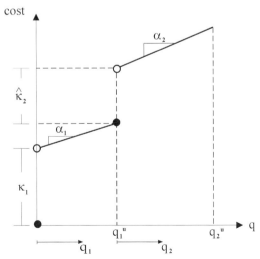

FIGURE 7.4 Variables used to define the piecewise linear cost function in the optimization formulation.

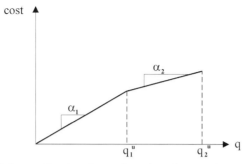

FIGURE 7.5 Unit operational costs can depend on the pumping regime. This scenario can be represented by piecewise linearization.

Finally, if q_1 is nonzero then (7.18) ensures that X_1 is one and q_1 is less than or equal to its upper bound. The combination of (7.18), (7.19), and (7.20) will force q_1 to be at its bound when both binary variables are active.

The approach described here can be extended to three or more stress segments and can be used to solve several related problems. If κ_1 is set to zero, then this approach could be used to represent the case of an existing well and pump that can operate up to a rate of q_1^u with operation beyond this point requiring additional construction costs. If both construction cost coefficients are removed, then the formulation reduces to simply minimizing the operational costs when the cost per unit stress changes with the stress rate. The resulting piecewise linear objective function would take the form shown in Figure 7.5. In this case, the binary variables would still be required to define the relationships between active stress segments.

7.3 SOLVING MIXED BINARY FORMULATIONS

A variety of solution algorithms are available for the mixed binary linear problem. In all cases, solution of this problem is more computationally expensive than solution of problems that do not contain binary variables. The branch and bound algorithm is the oldest and most commonly applied of those available and will be presented here to demonstrate that these problems can be solved in a robust fashion.

7.3.1 BRANCH AND BOUND ALGORITHM

The general mixed binary linear formulation can be rewritten in compact form as

$$\text{minimize} \quad f = \beta + \sum_{j=1}^{n} \alpha_j q_j + \sum_{i=1}^{l} \gamma_i h_i + \sum_{b=1}^{n_b} \kappa_b X_b \qquad (7.21)$$

such that
$$\sum_{j=1}^{n} \alpha_{j,k}^{q} q_j + \sum_{i=1}^{l} \alpha_{i,k}^{h} h_i + \sum_{b=1}^{n_b} \alpha_{b,k}^{X} X_b + \beta_k \geq 0,$$

$$\text{for } k = 1, \ldots, m \tag{7.22}$$

$$0 \leq X_b \leq 1 \qquad b = 1, \ldots, n_b \tag{7.23}$$

$$X_b \text{ is integer} \qquad b = 1, \ldots, n_b \tag{7.24}$$

Note that the formulation is a simple linear program if the integrality requirements on X_b are removed. If these constraints are removed and a feasible solution exists to the resulting linear problem, then two possible outcomes of decision variable values can arise. The first possibility is that all of the binary variables can happen to take binary values. The second, and more likely outcome, is that some of the binary variables take on non-binary values between zero and one. This outcome is, of course, not feasible with respect to the integrality constraints. In the first case, however, the integrality constraints are satisfied and the binary formulation will have been solved.

The branch and bound algorithm solves a series of problems where the integrality constraints are relaxed in search of a solution that satisfies, by happenstance or constraint, all integrality constraints. In each of the branch and bound problems, the value of some of the binary variables may be constrained to be either zero or one.

The branch and bound algorithm begins by solving the mixed binary program with integrality constraints relaxed. If there is no feasible solution to this problem, then the original mixed binary program will not have a feasible solution. Assuming that a feasible solution to the relaxed problem exists, the resulting values of the binary variables are examined. If all binary variables take on a binary value (either zero or one), then the solution is feasible with respect to the integrality constraints and the optimal solution has been obtained.

If at least one binary variable does not take on a binary value, then the algorithm proceeds. It is important to recognize that the objective function value obtained at this stage is the smallest value possible. That is, if all integrality constraints are enforced, the objective cannot be less than the value achieved when the constraints are relaxed. The next step is to select one of the binary variables whose value was not binary in the initial solution. Two subproblems are then defined and solved. In one of the subproblems the chosen binary variable is forced to take the value zero; in the other it is forced to take the value one. This forced assignment of values constitutes the bounding in the branch and bound algorithm, and the solution of subproblems constitutes the branching of the algorithm.

The values of the binary variables in each of the two subproblems are examined. If at least one of the binary variables does not take a binary

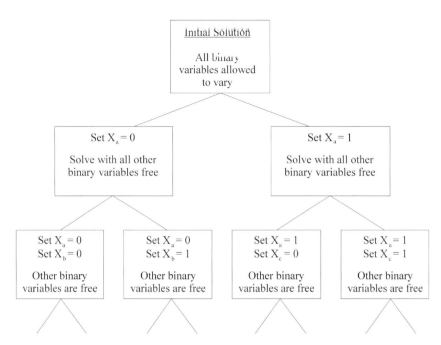

FIGURE 7.6 The branch and bound algorithm proceeds by constraining the values of specific integer variables and relaxing other integrality constraints.

value, then a new branching is required and two additional subproblems are solved, now with two of the binary variables fixed at specified values. The generation of the branching tree is depicted in Figure 7.6. Each box represents the solution of a linear program in which all but a subset of the binary variables are allowed to vary freely between zero and one. The subset of binary variables that are fixed is indicated. The subscripts a, b, and c indicate some binary variable that has been set. Note that in the second level of the right main branch a different binary variable (X_c) is selected than in the left main branch (X_b). This can arise because the setting of X_a to zero and one can produce two alternative solutions with different combinations of binary variables satisfying the integrality requirement.

Branching and bounding continue on each branch until a solution occurs in which all binary variables take binary values. In general, complete enumeration of every branch requires examining every combination of zero–one values for the binary variables. Complete enumeration would require a very large number of solutions when the number of binary variables is large. The branch and bound algorithm avoids an exhaustive search by utilizing a simple observation.

Suppose two solutions are available. In the first solution all binary variables take binary (zero–one) values. This will be called the binary solution. The second solution contains some binary variables that take nonbinary values. This will be called the nonbinary solution. Suppose that the objective function value for the nonbinary solution is larger than the objective function value of the binary solution. It can be concluded that further branching from the nonbinary solution can never yield a solution with a lower objective function value than the binary solution. This is proved by considering that placing further restrictions on the nonbinary solution can only increase the objective function value further.

The branch and bound strategy for enumeration is to descend the most promising branch until a binary solution is determined. This solution can then be compared with the solutions for other branches for rapid elimination of nonpromising branches. The most promising branch is that which follows the path of lowest objective function value at each branching step.

The branch and bound algorithm is guaranteed to converge. However, it may require substantial computational effort. The branch and bound algorithm can require solution of a large number of linear programs with the computational effort related to the number of binary variables that are used.

7.3.2 INTERPRETING SOLUTION RESULTS

Linear programs that include both continuous and integer variables in the objective function are analogous to the multiobjective problems discussed in Chapter 5. The relative magnitudes of the objective function coefficients on the decision variables will influence the solution. If the objective is to minimize the total cost, which includes linear operational costs and fixed charges, then the ratio of the unit costs is important in solving the problem. When the fixed cost is large relative to operational costs, then the problem becomes one of minimizing the number of locations and the stresses are secondary. Conversely, when operational costs dominate, the fixed costs do not significantly influence the solution. Sensitivity analysis should be performed to determine the sensitivity of model output to the ratio of these two costs. If one of the costs is inconsequential to the solution, then resources do not need to be expended to minimize the uncertainty inherent in that value.

The range analysis presented in Section 5.2 is not applicable for integer problems. That analysis is predicated on the assumption that all decision variables are continuous. The presence of integer variables violates this assumption. Nonetheless, the other sensitivity analysis procedures, which involve manipulating the formulation and resolving the problem, are still valid for analyzing mixed integer output.

7.3.3 THE WATER SUPPLY EXAMPLE

We return again to the water supply example previously discussed in Sections 3.7, 4.4, 5.1.2, and 5.2.1 and introduce a formulation that includes integer variables. Assume that for the water supply problem we need to determine the extraction rates from the two candidate locations shown in Figure 7.7. Furthermore, the supply wells are expected to operate for only 5 years as a temporary source of water. Rather than maximizing the total extraction, as we did in previous chapters, this time we seek to maximize the net proceeds from supplying the water over a 5-year time frame. The net proceeds are to include the net income from supplying the water minus the installation costs. This formulation is stated as

$$\text{maximize} \quad f(\mathbf{q}, \mathbf{X}) = \sum_{j=1}^{2} \alpha_j q_j - \sum_{j=1}^{2} \kappa_j X_j \qquad (7.25a)$$

such that

$$h_A \geq h_{\text{river}, A} \qquad (7.25b)$$

$$h_B \geq h_{\text{river}, B} \qquad (7.25c)$$

$$h_C \geq h_{\text{river}, C} \qquad (7.25d)$$

$$q_1 \leq M X_1 \qquad (7.25e)$$

$$q_2 \leq M X_2 \qquad (7.25f)$$

$$X_1 + X_2 \leq N^u \qquad (7.25g)$$

$$q_1^l \leq q_1 \leq q_1^u \qquad (7.25h)$$

$$q_2^l \leq q_2 \leq q_2^u \qquad (7.25i)$$

$$0 \leq X_1 \leq 1 \qquad (7.25j)$$

$$0 \leq X_2 \leq 1 \qquad (7.25k)$$

$$X_1, X_2 \text{ are integers} \qquad (7.25l)$$

The formulation is not complete until values for the coefficients α_j, κ_j, M, N^u, and the upper and lower bounds for the stresses are specified. The installation cost, κ_j, is incurred at the beginning of the project so that it is a present value cost. Well installation costs vary, depending on geology, well diameter, well depth, well construction material, and several other factors, but for the sake of this example, assume that the well installation cost is $75,000.

The net income from supplying a unit of water must be discounted over the life of the project, as discussed in Section 6.1.3 and expressed in equa-

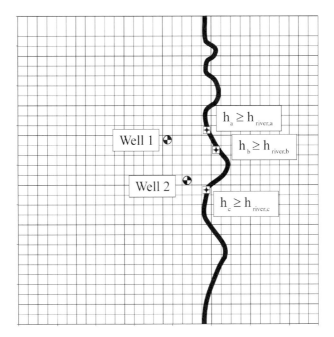

FIGURE 7.7 Numerical grid for the water supply example. Water supply wells and head constraint locations for the hydraulic control formulation are shown.

tion (6.7). Assume a billing rate by the water utility of $1 per 1000 gallons ($0.2642/m³) of water delivered, and further assume that the water utility realizes a 15% return so that the net income is $0.15 per 1000 gallons ($0.0396/m³). The present value of a constant annual stream of income can be determined from

$$P = A\frac{(1+i)^n - 1}{i(1+i)^n} \tag{7.26}$$

where P is the present value of an annual series, A is the annual net income, i is the interest rate, and n is the number of years. The net annual income is the product of the marginal income (i.e., the net income per unit stress rate) and the volume of water supplied over a year. In our example, the stresses are expressed in units of (m^3/day), so we also must include the proper conversion factors

$$A = \left(\frac{\$0.0396}{m^3}\right)\left(\frac{365 \text{ days}}{\text{year}}\right)q = 14.46\,q \tag{7.27}$$

where the number 14.46 has units of $(\$ \cdot d/m^3 \cdot \text{year})$. Assuming an 8% interest rate over the 5-year project, the present value of the net income is calculated as

$$P_j = (14.46\,q_j)\frac{(1.08)^5 - 1}{(0.08)(1.08)^5} = 57.73\,q_j \tag{7.28}$$

The present value per unit stress from supplying the water is α_j, the coefficient on q_j in equation (7.28).

The value M in constraints (7.25e) and (7.25f) must be sufficiently large that the constraints do not restrict the values of the stresses below some upper bound. Instead, the point of M is to ensure that the binary variables are set to either 1 or 0, depending on the value of q_j. Therefore, M must be at least as large as the upper bounds on the stresses; when M is set to the respective upper bounds, then the upper bounds in constraints (7.25h) and (7.25i) are redundant and can be eliminated. The upper bounds on stress may be established based on hydraulic considerations of the well design. Let us assume that for the costs just indicated, the well capacity is 10,000 m³/day. Although we can place nonzero lower bounds on the stress, we will leave the lower bounds at zero so that we can compare the results from the binary linear program with those of the continuous problem that is solved in Chapter 4.

The maximum number of active stress locations is controlled by the variable N^u in constraint (7.25g). With only two candidate locations, this parameter can take on only the value 0, 1, or 2. Of course, a value of 0 would make the formulation trivial because no wells would be allowed to operate; N^u must be either 1 or 2.

Combining all of the preceding coefficients, the full water supply formulation is stated as follows:

$$\text{maximize} \quad f(\mathbf{q}, \mathbf{X}) = (\$57.73) \sum_{j=1}^{2} q_j - (\$75,000) \sum_{j=1}^{2} X_j \tag{7.29a}$$

$$\text{such that} \quad h_A \geq h_{\text{river}, A} \tag{7.29b}$$

$$h_B \geq h_{\text{river}, B} \tag{7.29c}$$

$$h_C \geq h_{\text{river}, C} \tag{7.29d}$$

$$q_1 \leq (10,000)\, X_1 \tag{7.29e}$$

$$q_2 \leq (10,000)\, X_2 \tag{7.29f}$$

$$X_1 + X_2 \leq 2 \tag{7.29g}$$

$$q_1 \geq 0 \tag{7.29h}$$

$$q_2 \geq 0 \tag{7.29i}$$

$$0 \leq X_1 \leq 1 \tag{7.29j}$$

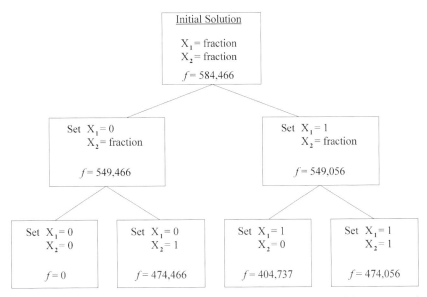

FIGURE 7.8   Branch and bound diagram for the binary formulation of the water supply problem.

$$0 \leq X_2 \leq 1 \tag{7.29k}$$

$$X_1, X_2 \text{ are integers} \tag{7.29l}$$

The branch and bound solution to this formulation is depicted in Figure 7.8. Note that at the first branch, the first binary variable is set to either 1 or 0 and the second binary variable is relaxed. In both subproblems in the first branch, X_2 is a fraction at the optimal solution; therefore branching must be done from both subproblems. In this simple example, full enumeration of all four possible binary solutions is required. For larger problems, this will generally not occur. Nevertheless, we see from Figure 7.8 that the optimal solution is to install well 2 but not to install well 1. The optimal pumping rate for well 2 is 9,528 m³/day. Compare this solution with the continuous variable formulation from Chapter 4, where the solution is to pump 2,393 m³/day from well 1 and 8,435 m³/day from well 2. The inclusion of installation costs in the integer formulation resulted in eliminating well 1 from the solution and led to a 12% reduction in water supply.

Comparing the objective function values in the last row of Figure 7.8, we see that there is only a small difference between the values of the optimal solution and the solution that requires both wells to operate. This indicates that the installation costs and the operational costs are both playing important roles in the optimal solution and that care must be taken in determining the two cost components. For example, if the installation costs are only $500 cheaper, then the solution is to pump both wells at the same rates as in the continuous formulation.

7.4 NOTES AND REFERENCES

A number of mixed binary formulations that follow the form presented in Section 7.1 have appeared in the literature. These include Aguado and Remson (1980), Lall and Santini (1989), and Dougherty and Marryott (1991), who solved a two-dimensional construction dewatering problem; Galeati and Gambolati (1988), who solved a three-dimensional dewatering model; Ritzel *et al.* (1994), who solved a plume containment problem; Sawyer *et al.* (1995), who solved a large-scale application to a three-dimensional plume control problem; and Misirli and Yazicigil (1997), who solved a water supply problem with simultaneous plume containment. All of these authors use the binary variable to add a fixed charge term to the objective function. Other methods for incorporating fixed charges that do not require binary variables are discussed in Chapter 8.

In a manner similar to that described in Section 7.2.5, Danskin and Gorelick (1985) use continuous variables in combination with binary variables to construct a piecewise linear function that represents the relationship between streamflow and recharge.

The method for relating the stress variables to the integer variables embodied in equation (7.2) is a standard approach drawn from the operations research literature and is described along with solution methods for integer programming problems (Garfinkel and Nemhauser, 1972; Nemhauser and Wolsey, 1988; Hillier and Lieberman, 1995).

8

FORMULATIONS WITH NONLINEAR FUNCTIONS

For many applications, the objective or constraint functions are nonlinear functions of the decision variables. Nonlinearities arise from two sources in groundwater management problems. The first source is a nonlinear response of system state to stress. The primary example of this nonlinearity is unconfined aquifers, where the head is a nonlinear function of stress. The second source of nonlinearity results from objective and constraint functions that depend nonlinearly on the system decision variables or system state.

If either of these nonlinear conditions exists, then the constraint or objective functions are nonlinear and alternative solution algorithms must be used. Generally, these algorithms will require more computational effort to obtain a solution. A further complication is that, in some cases, it cannot be guaranteed that the solution algorithm will find the optimal solution. In this chapter, we describe a number of formulations in which nonlinear functions arise and examine the potential for local minima. Several solution algorithms are introduced and discussed.

8.1 FEATURES OF NONLINEAR OPTIMIZATION PROBLEMS

A general statement of the groundwater management model encompassing all the forms introduced in previous chapters, including binary variables, will take the form

$$\text{minimize} \quad f(q_1, q_2, \ldots, q_n, X_1, X_2, \ldots, X_{n_b}, h_1, h_2, \ldots, h_l) \quad (8.1)$$

such that　$g_k(q_1, q_2, \ldots, q_n, X_1, X_2, \ldots, X_{n_{ij}}, h_1, h_2, \ldots, h_l) \leq 0$

$$k = 1, \ldots m$$

where f is the objective function and g_k is the kth constraint function. If f and g are required to be linear functions, then this form reduces to the general binary formulation given in Chapter 7. However, this form also accommodates the possibility that the objective or constraint functions are nonlinear functions of the decision variables.

A somewhat simpler formulation is one in which all variables are continuous, given as

$$\text{minimize}\quad f(q_1, q_2, \ldots, q_n, h_1, h_2, \ldots, h_l) \tag{8.2}$$

$$\text{such that}\quad g_k(q_1, q_2, \ldots, q_n, h_1, h_2, \ldots, h_l) \leq 0$$

$$k = 1, \ldots m$$

The introduction of nonlinear functions adds complexity both to the methods used to solve the resulting optimization problem and to the interpretation of the results. In this section, some of the characteristics of nonlinear functions and nonlinear optimization problems are reviewed. For purposes of this review consider a single function, f, of a single continuous variable, x.

8.1.1 FUNCTION CONTINUITY

The first distinction to make is between continuous and discontinuous nonlinear functions. The function f is continuous at a given value of x if the following condition is true:

$$\lim_{h \to 0} f(x + h) = f(x) \tag{8.3}$$

A function is continuous if this condition holds for all values of x. Figure 8.1a depicts a continuous function while 8.1b depicts a function that is discontinuous at the point x_a. The multiple segment objective functions introduced in Section 7.2.5 are examples of discontinuous functions when considered with respect to the single stress.

8.1.2 FUNCTION CONVEXITY AND CONCAVITY

An important property of a nonlinear, continuous function is its "shape." An important measure of a function's shape is its convexity or concavity. A continuous function f is convex if

$$f(\omega x_a + (1 - \omega)x_b) \leq \omega f(x_a) + (1 - \omega)f(x_b) \tag{8.4}$$

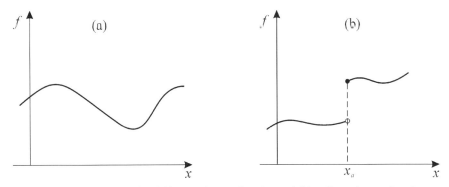

FIGURE 8.1   Example of (a) a continuous function and (b) a discontinuous function.

for any pair of values x_a and x_b and for $0 \le \omega \le 1$. This statement implies that a line can be drawn connecting the function points evaluated at x_a and x_b that will always be above the function at all points between x_a and x_b. This concept is depicted in Figure 8.2 for a convex function.

A concave function has properties opposite to those of a convex function. A function is concave if

$$f\big(\omega x_a + (1 - \omega)x_b\big) \ge \omega f(x_a) + (1 - \omega)f(x_b) \tag{8.5}$$

for any pair of values x_a and x_b and for $0 \le \omega \le 1$. In this case, the function must always lie above the line as depicted in Figure 8.3. Another way to define concavity is to say that the function f is convex if $-f$ is concave.

A common occurrence for nonlinear continuous functions is that the function is convex or concave only in portions of the range of independent

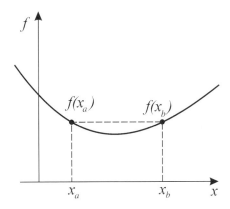

FIGURE 8.2   Example of a convex function.

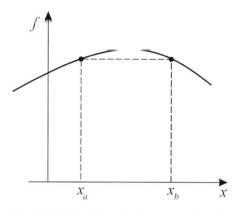

FIGURE 8.3   Example of a concave function.

variable values. Figure 8.4 depicts such an example, where the function is convex in the region between x_a and x_b and concave in the region between x_b and x_c. Such a function is referred to as nonconvex or nonconcave; that is, over the entire range of independent variable values the function is neither convex nor concave.

A final observation about convexity and concavity relates to linear functions. A linear function is a special case because it satisfies, as an equality, both the convexity and the concavity conditions. Hence, a linear function is both convex and concave.

The definitions provided in (8.4) and (8.5) illustrate the differences between concavity and convexity. However, using these definitions to determine which property applies to a given function requires function evaluation at numerous points, which is generally not practical. In order to

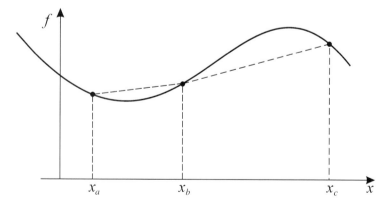

FIGURE 8.4   A nonconvex function.

determine whether a function is concave or convex, we turn to the derivatives of the function. A function of a single variable is convex in the range x_a to x_b if

$$\frac{\partial^2 f(x)}{\partial x^2} \geq 0 \qquad \text{for } x_a \leq x \leq x_b \qquad (8.6)$$

Conversely, a function is concave if the second derivative is less than or equal to zero. Note again that a linear function will always satisfy both the concavity and convexity conditions as an equality because the second derivative of a linear function is everywhere zero.

The concepts of convexity and concavity can be extended to multiple dimensions. The definitions provided by (8.4) and (8.5) are extended by evaluating the function at locations $\mathbf{x}_a$ and $\mathbf{x}_b$ in multidimensional space and connecting these points by a vector. The derivative-based definition is extended by considering a multidimensional derivative. A matrix of second derivatives, called the Hessian matrix, consists of the elements

$$\frac{\partial^2 f(\mathbf{x})}{\partial x_i \partial x_j} \qquad (8.7)$$

If the Hessian is positive semidefinite for all feasible values of the independent variables, then the function is convex over the feasible region. Linear functions have zero Hessian matrices and satisfy this condition. For nonlinear functions, it is sufficient to require that the Hessian be positive definite. A positive definite matrix has the properties that the determinant of all the matrices down the main diagonal is positive and all of its eigenvalues are positive. A sufficient condition for positive eigenvalues of the Hessian matrix is that the diagonal elements of the Hessian are positive and the sum of absolute values of elements in any row, excluding the diagonal, is less than the diagonal value.

The constraint functions, taken together, form a feasible region. Such a region is depicted in Figure 4.1 for a problem where the constraint functions are linear. The shape of the feasible region formed by nonlinear constraint functions significantly affects solution of the problem. A feasible region is convex if for any two points $\mathbf{x}_a$ and $\mathbf{x}_b$ that are contained in the feasible region the linear combination of the two

$$\omega \mathbf{x}_a + (1 - \omega)\mathbf{x}_b \qquad (8.8)$$

is also contained in the feasible region for $0 \leq \omega \leq 1$. This definition states that a line drawn between any two points in the feasible region must lie entirely within the feasible region. Figure 8.5a depicts a convex feasible region formed from the indicated constraint functions. By inspection, a line connecting any two points in the feasible region remains within the region. A counterexample is offered by Figure 8.5b. Here, the line connecting $\mathbf{x}_a$ and $\mathbf{x}_b$ leaves the feasible region. Hence, this is a nonconvex feasible region.

a

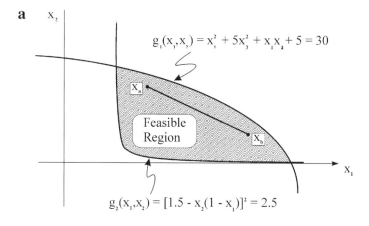

$$g_1(x_1,x_2) = x_1^2 + 5x_2^2 + x_1x_2 + 5 = 30$$

Feasible Region

$$g_2(x_1,x_2) = [1.5 - x_1(1 - x_2)]^2 = 2.5$$

b

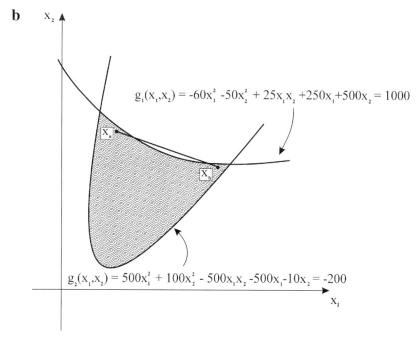

$$g_1(x_1,x_2) = -60x_1^2 - 50x_2^2 + 25x_1x_2 + 250x_1 + 500x_2 = 1000$$

$$g_2(x_1,x_2) = 500x_1^2 + 100x_2^2 - 500x_1x_2 - 500x_1 - 10x_2 = -200$$

FIGURE 8.5   A convex feasible region is shown in (a) and a nonconvex feasible region is depicted in (b).

Some practical guidelines for determining the convexity of feasible regions can be drawn from the convexity properties of functions. The set of points that satisfy a constraint function define a feasible region for that constraint. If the constraint function is of the form $g(x) \leq 0$, that is, it is upper bounded by zero, and the constraint function is convex, then the feasible region for that constraint is also convex. When multiple constraints are present, the feasible region, which is constructed from the intersection of the feasible regions for the individual constraints, will be convex if all individual constraint feasible regions are convex. It follows that a feasible region will be convex if it is defined by convex, upper-bounded constraint functions. Another useful property of convex functions is that the sum of two convex functions is also convex. Considering the case of linear functions, we can note that the general linear formulation posed in (4.1) represents the minimization of a convex function on a convex feasible region.

8.1.3 OPTIMALITY CONDITIONS AND CONVEXITY

The convexity, or lack thereof, of an optimization formulation has implications for both the algorithm selected to solve the problem and the interpretation of the results. Consider the problem of minimizing a single function of a single variable. An example function is depicted in Figure 8.6. Minimizing this function implies finding the value of x with the lowest function value. In the range depicted, this global minimum occurs at x_1. However, the function has a second local minimum at x_2. A local minimum is a point that has a function value smaller than that found at any other point in a small region around the local minimum. A nonlinear function may have multiple local minima. The global minimum is one or more of the local minima that have the lowest function value. Note that if multiple points have the same globally minimum function value then each is a global minimum.

The presence of local minima has implications for the performance and results obtained by optimization algorithms. Nonlinear optimization algorithms are based on seeking a minimum function value. However, these

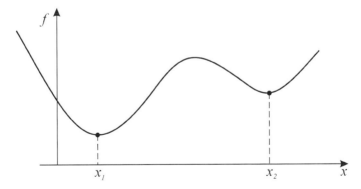

FIGURE 8.6   A function with two local minima, at x_1 and x_2, and one global minimum, at x_1.

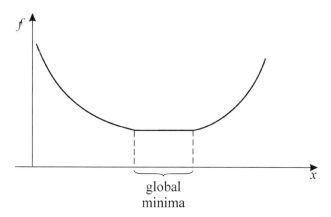

global
minima

FIGURE 8.7   A convex function with multiple local and global minima.

algorithms have difficulty determining whether a minimum is a local or global minimum. At present, no practical algorithm exists that can reliably determine the global minimum for problems of significant size.

Convexity can be used to determine the potential for the presence of local minima through the observation that, if a function is convex, then all local minima are global minima. Figure 8.7 depicts a function that is convex but has multiple local minima, all of which are also global minima. A more common case is that in which a single minimum exists. For a convex function, a single minimum will be both a local and a global minimum. It is important to note that this observation does not imply that if a function is nonconvex it has multiple local minima. Figure 8.8 provides an example of a function that is not convex yet has a single minimum.

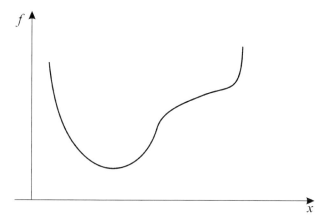

FIGURE 8.8   A nonconvex function with a single minimum.

The utility of convexity for predicting the potential for multiple local minima can be extended to multidimensional problems and to constrained problems. A useful general statement is that if a problem consists of minimizing a convex objective function subject to a convex feasible region then all local minima are global minima.

The determination of convexity or nonconvexity provides a powerful tool for interpreting the results of an optimization algorithm. As noted in Section 3.2.4, the general linear groundwater flow management problem consists of a linear (convex) objective and linear (convex) constraints that form a convex feasible region. Hence, the simplex algorithm, which determines the solution to the linear optimization problem, is guaranteed to find the global minimum to the problem. In some cases the convexity of nonlinear problems can also be proved.

8.2 EXAMPLES OF NONLINEAR FORMULATIONS

Nonlinear problems arise when either the objective or constraint functions are nonlinear functions of the decision variables. The possible forms of nonlinear functions are virtually unlimited. In this section, we describe a series of nonlinear formulation elements that are in common use and which serve to demonstrate the potential implementation of nonlinear functions. For each nonlinear function, an exploration is made of the nature of the nonlinearity and the presence or absence of convexity.

8.2.1 STRESS COSTS

A term that might be considered for inclusion in the objective function is the operational cost of applying stresses. The cost of pumping groundwater and conveying it through a piping system can be related to several factors, one of which is the cost of electrical power. As an example of such a cost function, consider energy costs associated with lifting water from the aquifer to the ground surface. The problem is arranged as shown in Figure 8.9 for a confined aquifer. To lift the water from the well an amount of energy, equivalent to the change in head $(L - h)$, must be added to the fluid, where h, the head in the well, depends on the pumping rate.

The power requirements to lift the water can be determined by applying the conservation of energy equation to this problem. This yields

$$gL = gh + w - gh_L \tag{8.9}$$

where w is the power input to the fluid per unit mass flux, h_L is the head loss due to pipe friction involved in moving the water from a head h to

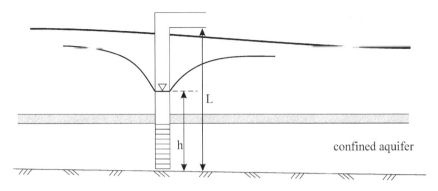

FIGURE 8.9   Example of incorporating lift costs.

L, and g is the gravitational constant. Rearranging, and introducing ρ, the fluid density, and q, the flow rate, the power requirement is

$$W = \rho g q (L + h_L - h) \qquad (8.10)$$

Defining H to be $(L + h_L)$ and introducing a positive coefficient, C, which may include pump efficiency, the power requirements for well k can be written as

$$W_k = C_k q_k (H_k - h_k) \qquad (8.11)$$

Note that H might also be defined as the head to which the water must be raised to drive the water through a surface piping system to some terminal point. Whenever H includes pipe flow head losses it is possible that H will be a function of q.

With this definition of pumping power, the total pumping power for the system can be minimized as

$$\text{minimize} \quad W = \sum_{k=1}^{n} W_k = \sum_{k=1}^{n} C_k q_k (H_k - h_k) \qquad (8.12)$$

In its most complete form, (8.12) is a nonlinear function of pumping rates because both H and h depend on the pump rate. One level of simplification is to specify that H is a constant. This presumes that either the head loss is small compared with L or that the head loss varies little over the range of flow rates that are expected to be considered by the optimizer. In some circumstances, changes in h as a result of pumping are small compared with $(H - h)$, in which case it might be reasonable to assume that h is constant. Combining the assumptions that h and H are constant, the power requirement reduces to a linear function of pumping rates.

If H is assumed constant but the dependence of h on q is retained, a nonlinearity arises. For a confined aquifer, h is a linear function of pumping and W is a quadratic function. This form can be derived by substituting the first-order Taylor series (4.4) for h_k in (8.12), which yields

$$W = \sum_{k=1}^{n}\left(C_k\left[(H_k - h_k^0)q_k - \left(\sum_{j=1}^{n}\frac{\partial h_k}{\partial q_j}q_j\right)q_k\right]\right) \tag{8.13}$$

Analysis of the derivatives of (8.13) reveals that this objective function may be convex under many circumstances. The power cost objective function (8.13) is quadratic and therefore the Hessian terms are constants because the change in head with respect to pumping is constant. These terms take the form

$$\frac{\partial^2 W}{\partial q_j \partial q_k} = -\alpha C_k \frac{\partial h_k}{\partial q_j} \tag{8.14}$$

where α is 2 if $k = j$ and 1 if $k \neq j$. Recalling that a positive value of q implies extraction, for nearly every conceivable case the derivative will be negative because increases in extraction produce decreases in head. It follows that the diagonal elements of the Hessian will be positive. It can also be observed that the second derivative of power with respect to pumping will probably be largest when $k = j$. Hence, the diagonal element will probably be much larger than the off-diagonal elements of the Hessian. If the diagonal elements are positive and dominate the row sums, then the function is convex.

Accounting for the dependence of both h and H on q adds a complex nonlinearity. Because H is a function of head losses in the piping system, the functional form may be quite complicated and depends on the type of head loss representation used. Further complexity is added if head losses are considered for pipelines that combine flows from multiple wells. Analysis can proceed by considering a common means of expressing head loss:

$$h_L = K_f q^2 \tag{8.15}$$

where K_f includes terms describing the geometry of the pipe and related elements, roughness characteristics, and other constants. Assuming that K_f does not depend on flow rate and that head losses from combined flows from individual wells are not included, (8.13) can be rewritten as

$$W = \sum_{k=1}^{n}\left(C_k\left[((L + K_f q_k^2) - h_k^0)q_k - \left(\sum_{j=1}^{n}\frac{\partial h_k}{\partial q_j}q_j\right)q_k\right]\right) \tag{8.16}$$

or

$$W = \sum_{k=1}^{n}\left(C_k\left[(L - h_k^0)q_k - \left(\sum_{j=1}^{n}\frac{\partial h_k}{\partial q_j}q_j\right)q_k + K_f q_k^3\right]\right) \tag{8.17}$$

The objective function is now cubic. The j, k term of the Hessian matrix takes the form

$$\frac{\partial^2 W}{\partial q_j \partial q_k} = -\alpha C_k \frac{\partial h_k}{\partial q_j} + \beta 6 K_f q_k \qquad (8.18)$$

where β is 1 if $k = j$ and 0 if $k \neq j$. Assuming q_k is positive, (8.17) is even more likely to be convex than (8.13) because of the additional positive term added to the diagonal in (8.18).

8.2.2 UNCONFINED AQUIFERS

Unconfined aquifers are defined by the governing equations described in Chapter 2. Because these equations involve a product of the independent variable and its derivative, they are nonlinear differential equations. As noted in Chapter 2, the numerical solution of the unconfined flow equations also produces a nonlinear system of equations. A physical interpretation of this nonlinearity is possible by considering that solving the unconfined problem requires simultaneously determining the hydraulic head and the domain of the problem. For three-dimensional problems this means determining the location of the free surface. For two-dimensional (areal) problems, determining the location of the free surface implies finding the thickness of the aquifer.

Because the solution domain is dependent on the head, these problems are nonlinear with respect to stress. If an optimization formulation contains hydraulic heads that are simulated using the unconfined flow equations, then a nonlinearity is introduced. Even if the heads appear in the formulation in linear functions, the objective or constraints that contain the heads will be a nonlinear function of the stress decision variable. The higher order terms of the Taylor series representation of hydraulic head as a function of stress (4.3) are no longer zero and the first-order Taylor series approximation is no longer exact.

Fully characterizing the presence or absence of convexity in unconfined problems is difficult. However, insight can be gained by considering the Dupuit–Forchheimer equation for radial flow to a fully penetrating well in a homogeneous unconfined aquifer. Figure 8.10 depicts the physical arrangement assumed for this solution with hydraulic conductivity, K, and pumping rate, Q

$$h^2 = H^2 + \frac{\ln\left(\frac{r}{R}\right)}{\pi K} Q \qquad (8.19)$$

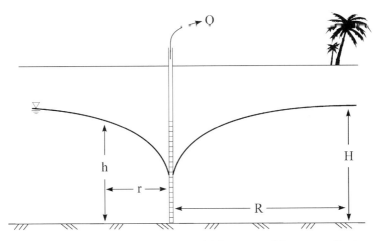

FIGURE 8.10 Physical arrangement for radial flow to a well in an unconfined aquifer.

where H is the specified reference head at a distance R from the pumping well and h is the head at a distance r from the pumping well. Equation (8.19) can be rearranged as

$$h = \sqrt{H^2 + \frac{\ln\left(\frac{r}{R}\right)}{\pi K} Q} \qquad (8.20)$$

to emphasize that head is related to the square root of pumping. Figure 8.11 provides a plot of head as a function of stress for K equal to 30 meters/day, (r/R) equal to 0.1, and H equal to 30 meters. The behavior is concave.

Examination of the second derivative of (8.20) shows that concavity is a general result for the Dupuit–Forchheimer solution. The second derivative is

$$\frac{\partial^2 h}{\partial Q^2} = -\frac{1}{4}\left(\frac{\ln\left(\frac{r}{R}\right)}{\pi K}\right)^2 \left[H^2 + \frac{\ln\left(\frac{r}{R}\right)}{\pi K} Q\right]^{-\frac{3}{2}} \qquad (8.21)$$

Note that the squared logarithmic term will always be positive. Similarly, the last term is simply $\left(h^2\right)^{-\frac{3}{2}}$, which will always be positive. It follows that the second derivative will always be negative and that the unconfined head is a nonconvex (concave) function of pumping rate.

A physical interpretation for this concavity can be provided by considering that an increase in pumping produces both an increase in gradient and a reduction in the saturated thickness through which water can flow. For example, when the pumping rate doubles the hydraulic gradient must more than double to account for the reduction in cross-sectional area through which flow occurs.

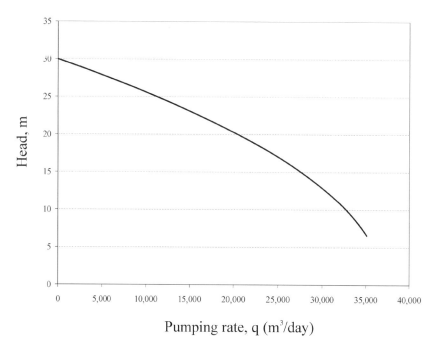

FIGURE 8.11   In unconfined aquifers, head is a nonlinear function of pumping.

Extending the conclusion of function concavity to problems with hydro-geologic complexity, multiple stress locations, and numerical solutions is not directly possible in part because linear superposition is not valid in the unconfined case. However, a few observations can be made. For the case of a single head with multiple stresses, it is likely that concave behavior will be present because the impact of multiple stresses will require the head gradient to increase faster than the pumping rate increases.

If it is assumed that hydraulic head is always a concave function of stress, then several additional observations can be made. In general, the sum of concave functions is also concave. For the case of multiple heads present in a single function (objective or constraint), the constraint or objective functions will be concave.

For some constraints a convex feasible region may be present. Consider a lower bound on hydraulic head, which can be written as

$$g(h_i) = h_i^l - h_i \leq 0 \qquad (8.22)$$

where h_i^l is a specified lower bound. If we assume that h_i is a concave function, then multiplying the function by negative 1 produces a convex function and the constraint function is convex. By similar reasoning and assumptions, an upper bound on head is likely to define a concave feasible region. Considering head difference constraints rewritten as

$$g\left(h_{k_1}, h_{k_2}\right) = -h_{k_1} + h_{k_2} + h_k^d \leq 0 \qquad (8.23)$$

we find a convex function added to a concave function. One or the other of these functions may dominate.

8.2.3 CONSTRUCTION COST AS A NONLINEAR FUNCTION

Figure 8.12a depicts the combined construction and operating cost function as discussed in Section 7.2.1. This function is discontinuous and can be represented using binary variables as described in Chapter 7. As long as the constraints and objective function are linear, the use of binary variables and the associated branch and bound algorithms is generally desirable because of the guarantee of a global solution and the robustness of the algorithms. However, if any of the elements of the problem are nonlinear, then the inclusion of binary variables produces a nonlinear binary problem, which is difficult to solve.

When other nonlinear elements are present it may be desirable to accommodate construction costs through a continuous nonlinear term in the objective function. This involves approximating the discontinuous function in Figure 8.12a with a continuous function that has a shape similar to that shown in Figure 8.12b. Such a function can be constructed using an exponential form

$$f = \alpha q + \kappa(1 - e^{-\beta q}) \qquad (8.24)$$

or a polynomial form

$$f = \alpha q + \kappa \left(\frac{q}{q + \beta}\right) \qquad (8.25)$$

where β is a large positive number in (8.24) and a small positive number in (8.25). For both functions, when q is zero the function is also zero. When q is much larger than zero the second term in the function approaches κ. As suggested by comparing Figure 8.12a and b, an error in the cost function of increasing magnitude occurs as q approaches zero. The amount of error can be controlled by selection of β.

By inspection of the functional form depicted in Figure 8.12b, it is clear that (8.24) is a concave function of q. This is confirmed by considering the computed second derivatives. The second derivative of (8.24) is

$$\frac{\partial^2 f}{\partial q^2} = -\beta^2 \kappa e^{-\beta q} \qquad (8.26)$$

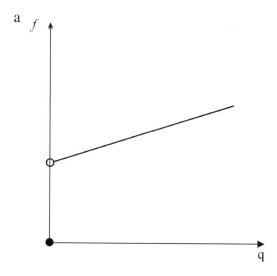

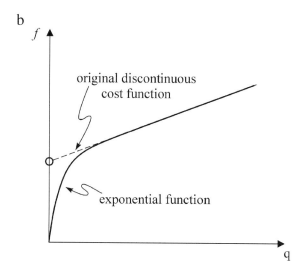

FIGURE 8.12   A discontinuous function (a) can be approximated with a continuous, concave function (b).

and will always be negative. Similarly the second derivative of (8.25) is

$$\frac{\partial^2 f}{\partial q^2} = -\frac{2\kappa\beta}{(q+\beta)^3} \tag{8.27}$$

which will also always be negative because β is always positive.

8.2.1 FLUX BETWEEN AQUIFER AND STREAM

The dependence of flux between aquifer and stream on aquifer head is described in Section 6.5 and summarized here as

$$q_{\text{stream}} = C(h_i - h_{\text{stream}}) \qquad \text{if } h_i \geq b_{\text{stream}}$$
$$q_{\text{stream}} = C_0(b_{\text{stream}} - h_{\text{stream}}) \qquad \text{if } h_i < b_{\text{stream}} \tag{8.28}$$

Cases may arise in which it is desirable to place constraints on q_{stream}. This might be necessary if stream recharge is to be controlled in the vicinity of water supply wells. However, it is difficult to constrain q_{stream} when its functional form depends on the hydraulic regime.

One approach to this problem is to establish a series of constraints. The first two constraints define two possible values of the stream flux

$$q_h = C(h_i - h_{\text{stream}})$$
$$q_b = C_0(b_{\text{stream}} - h_{\text{stream}}) \tag{8.29}$$

Note that if $h_i < b_{\text{stream}}$ then q_h will be less than q_b. Similarly, q_h will be greater than q_b when the aquifer head is above the stream bed. Hence, the variable q_{stream} should be set to the maximum of these two possible values. This is accomplished by requiring that the stream flux be greater than or equal to either of the two possible values

$$q_{\text{stream}} \geq q_h$$
$$q_{\text{stream}} \geq q_b \tag{8.30}$$

and that the product of the differences between possible and final stream fluxes be zero

$$(q_{\text{stream}} - q_h)(q_{\text{stream}} - q_b) = 0 \tag{8.31}$$

The product constraint (8.31) forces the stream flux to be equal to one or the other of the two possible fluxes, while the upper bound constraints (8.30) force the stream flux to be equal to the larger of the two possible fluxes. The product constraint introduces a nonlinear constraint function through the products of q_{stream} with itself and with q_h. The necessity of computing q_{stream} in this manner arises when it is a decision variable in the formulation. Whereas q_b is a constant independent of stress, q_h depends on aquifer head, which in turn depends on stress.

The same problem can be addressed using binary variables to transform the discrete flux definition into a continuous function. First, define a binary variable, X_1, that indicates whether or not the stream is hydraulically connected to the aquifer. A single-equation definition of q_{stream} can then be posed as

$$q_{\text{stream}} = C(h_{\text{stream}} - h_i)X_1 + C_0(b_{\text{stream}} - h_{\text{stream}})(1 - X_1) \tag{8.32}$$

It now remains to create additional constraints that properly set the value of the binary variable. This is accomplished by

$$h_i + M \geq X_1 (b_{\text{stream}} + M)$$

$$b_{\text{stream}} + M \geq (1 - X_1)(h_i + M) \tag{8.33}$$

The first constraint forces X_1 to be zero when b_{stream} is greater than h_i. When b_{stream} is less than h_i, the second constraint forces X_1 to be one. Note that M (a large number) is required to ensure that the left side of the inequality is positive when the problem datum results in negative values for h_i or b_{stream}. The set of constraints contain nonlinearities produced by the product of the binary variables and the hydraulic heads.

8.2.5 MULTIPLE-CELL FACILITY LENGTH DETERMINATION

Multiple-cell facilities have been introduced and discussed in Section 6.3. The decision variables for these facilities are the location of the starting cell, the orientation of the facility, the length of the facility, and the total stress applied to the facility. As an additional example of a problem in which binary variables are present in nonlinear combination with continuous variables, consider the case in which the location of the starting cell of the facility and the orientation of the facility are presumed known, but the length of the facility and the total stress for the facility need to be determined.

As an example, consider a long vertical well that may extend over a maximum of n_w numerical layers. The design question is to determine the number of layers through which the vertical well will extend and the pumping rate in each layer. The binary variable X_j is used to indicate whether the facility extends to the jth layer, where the binary variables are numbered from 1 to n_w and X_1 refers to the starting location of the facility. Stress variables are also assigned along the length of the facility and identified by $q_j, j = 1, \ldots, n_w$. The relationship between stress and indicator variables is defined by (7.2), which requires that the indicator variable must be nonzero if the corresponding cell stress is nonzero. An additional requirement must be imposed to ensure that all indicator variables are nonzero from the starting location to the furthest active cell. (For example, constructing the facility to the jth cell requires constructing and activating all lower indexed cells, 1 through $j - 1$.) This requirement is satisfied by imposing

$$X_j \geq X_{j+1} \qquad j = 1, \ldots, n_w - 1 \tag{8.34}$$

This series of $(n_w - 1)$ constraints ensures that the indicator variable for all cells above any active cell will be set to one. The use of this constraint is not intended to imply that all cells with active indicator variables must also have nonzero cell stress but only that the facility will be constructed through this cell.

Formulations involving multiple-cell facilities do not meaningfully combine with constraints of the form of (7.5) and (7.7) because these constraints depend on the indicator variable being zero when the corresponding stress is zero. Conversely, the objective function must include a term that minimizes costs associated with construction. Such a term would take the form of a cost for construction through each cell along the path of the facility. Without such a term in the objective there exists no incentive in the formulation to set indicator variables to one only when they are needed.

In Section 7.2.4 several methods are introduced to ensure that the individual cell rates selected by the formulation are related to each other so as to mimic the actual behavior of the aquifer in response to the facility. The imposition of such a requirement here is complicated by the lack of prior knowledge of which cells will be active. One means of accomplishing this is to impose the requirement that

$$X_{j+1}h_j = X_{j+1}h_{j+1} \qquad j = 1, \dots, n_w - 1 \tag{8.35}$$

where h_j is the head at the jth cell of the facility. When the $j + 1$ cell is inactive, then X_{j+1} is zero and (8.35) is trivially satisfied. When the $j + 1$ cell is active then (8.35) imposes the same requirement as described in (6.14). The disadvantage of using (8.35) is that it introduces a nonlinear constraint.

8.3 SOLVING NONLINEAR FORMULATIONS

In Chapter 4, the simplex algorithm for solving linear, continuous variable problems is presented. The properties of the linear optimization problem are such that this single algorithm can successfully solve nearly any problem. Although research has identified alternative algorithms, the simplex algorithm remains the favorite method for solving the linear optimization problem.

In contrast, there exists a large suite of nonlinear optimization algorithms that are currently in use. No single algorithm is most effective for all classes of nonlinear optimization problems. A particular class of problems may be best solved by an algorithm that may not perform effectively on other classes of problems. The selection of an appropriate algorithm is tailored to the nature of the problem to be solved. Research on new algorithms and the relative effectiveness of existing algorithms is ongoing.

In this section, an overview of optimization algorithms relevant to groundwater management problems is provided. These techniques include sequential linearization methods, general gradient-based methods, and nongradient methods. A discussion of solving mixed nonlinear–binary problems is also included.

8.3.1 SEQUENTIAL LINEARIZATION FOR UNCONFINED PROBLEMS

The sequential linearization approach is applicable to the solution of problems in which the hydraulic head is simulated with unconfined equations but all other elements of the formulation are linear and no binary variables are present. Under these circumstances, an algorithm can be used that exploits the weak nonlinearity of head with respect to stress. Because head is only mildly nonlinear, an algorithm is posed that solves the problem as a series of linear programs.

The solution of the general linear formulation using the simplex algorithm is predicated on the linearity of hydraulic head response to stress. This implies that the derivative of head with stress is a constant. For the unconfined case the nonlinearity is mild and the derivative is nearly constant. This can be shown by examining the derivative of the head predicted by the unconfined equation with respect to stress. Recall (8.19), which relates aquifer head to pumping from one well. Taking the derivative of

$$h = \sqrt{H^2 + \frac{\ln\left(\frac{r}{R}\right)}{\pi K} Q} \qquad (8.36)$$

with respect to Q yields

$$\frac{\partial h}{\partial Q} = \frac{1}{2}\left(\frac{\ln\left(\frac{r}{R}\right)}{\pi K}\right)\left[H^2 + \frac{\ln\left(\frac{r}{R}\right)}{\pi K} Q\right]^{-\frac{1}{2}} \qquad (8.37)$$

Examination of (8.37) shows that when $H^2 \gg \left|\frac{\ln(r/R)}{\pi K} Q\right|$ then the derivative is nearly a constant. Because the magnitude of the logarithmic term is related to drawdown, $\left(\frac{\ln(r/R)}{\pi K} Q\right) = h^2 - H^2$, this condition implies that if the drawdown is much smaller than the thickness of the aquifer under low-stress conditions then the gradient is nearly constant and hence the head response is nearly linear. The behavior is suggested by the example depicted in Figure 8.11, where at low stresses (small drawdown) the gradient is nearly constant.

For each linear program the aquifer domain or thickness is assumed known and a response matrix is constructed in a manner similar to that for the confined problem. In effect, the aquifer is assumed to be confined in each iteration and the resulting linearly constrained problem is solved. The stresses are then used to simulate flow under unconfined conditions. The resulting heads are used to re-estimate the aquifer domain and the algorithm repeats with computation of a new response matrix. This can be written as an algorithm with iteration counter k.

SEQUENTIAL LINEARIZATION ALGORITHM

Step 0: Assume an initial aquifer domain, set $k = 1$, set $\mathbf{q}_0 = 0$.
Step 1: With fixed aquifer domain, construct a linear response matrix.
Step 2: Solve the linear program to determine optimal stresses, $\mathbf{q}_k$.
Step 3: If $\|\mathbf{q}_k - \mathbf{q}_{k-1}\| \leq \epsilon$ stop; else go to step 4.
Step 4: Simulate the heads produced by $\mathbf{q}_k$ assuming an unconfined aquifer.
Step 5: Use the heads predicted in step 4 to revise the aquifer domain.
Step 6: Set $k = k + 1$, go to step 1.

The algorithm can be viewed as sequentially linearizing the nonlinear head response to stress. This concept is depicted in Figure 8.13, where the stress is considered at three different points. At each point the derivative of the head with respect to stress provides the slope of the tangent line to the head response curve. At each iteration of the sequential linearization algorithm, the head response is assumed to follow the tangent line.

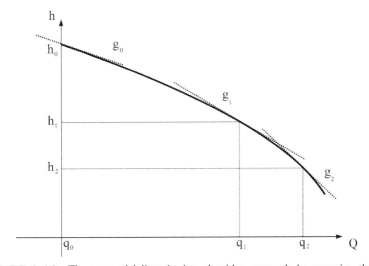

FIGURE 8.13 The sequential linearization algorithm proceeds by assuming that the aquifer domain is fixed at each iteration. The response coefficients, which represent the change in head with respect to pumping, are determined at the beginning of each iteration and held constant for that iteration.

Computation of the response matrix at each iteration is complicated by the nonlinear response. Recall that the Taylor series is used to project the value of head from a nominal value of stress to some other stress value. In the case of a confined aquifer, where the derivative of head with respect to stress is a constant, the first-order Taylor series is exact. This implies that the point from which the projection is made can be arbitrary. It is most convenient to make this point zero stress. In the unconfined case, the projection could again be made from the zero-stress point; however, accuracy of the projection might be substantially affected. An alternative, and probably superior approach, is to locate the point of projection as close to the solution as possible. The best estimate of this is the value of stress obtained from the previous iteration.

Returning to the general form of the Taylor series (2.17), a first-order Taylor series projection is constructed from the current solution to yield an approximation of the form

$$h_i^k(\mathbf{q}_{k+1}) = h_i^k(\mathbf{q}_k) + \sum_{j=1}^{n} \frac{\partial h_i^k}{\partial q_j}(\mathbf{q}_k)(q_j^{k+1} - q_j^k) \qquad (8.38)$$

where the superscript on the head variables indicates that they are computed using the aquifer domain from the kth iteration. Using this form of the Taylor series requires modifying the right-hand side of the constraint equations so that, for example, bound constraints at location i would be written as

$$\sum_{j=1}^{n} \frac{\partial h_i^k}{\partial q_j}(\mathbf{q}_k) q_j^{k+1} \leq h_i^u - h_i^k(\mathbf{q}_k) + \sum_{j=1}^{n} \frac{\partial h_i^k}{\partial q_j}(\mathbf{q}_k) q_j^k \qquad (8.39)$$

The nonlinearity of head response also affects the way in which the response coefficients are computed. Recalling the finite difference error analysis described in Section 2.3.3, a basic observation for nonlinear functions is that the larger the perturbation the less accurate is the difference approximation. For the confined case, an arbitrarily large perturbation value can be used due to the linear response of head. In contrast, in the unconfined case, the perturbation value must be selected with care. An excessive perturbation value could dewater the modeled aquifer so that no meaningful solution is possible. At the same time, excessive round-off error can cause the response coefficients to have inadequate precision when the perturbation is too small.

When implementing a sequential linearization algorithm, care must be taken while using the response coefficients, which represent a confined aquifer assumption. The derivative projection tends to overpredict the pumping required to meet the head bound for an unconfined aquifer (see Figure 8.13). This can lead to dewatering the aquifer at the start of the next iteration. It is sometimes necessary to scale back the pumping rate from the previous iteration in order to avoid dewatering the aquifer. A method for preventing aquifer dewatering is implemented in the MODOFC code and described in the accompanying documentation found in the CD-ROM attached to this book.

8.3.2 GRADIENT-BASED METHODS

The derivative of a function yields important information about the location of a minimum point of that function. For example, the function depicted in Figure 8.14 has a minimum at the point x_a. The derivative of the function at this point is zero. At point x_b the derivative is not zero, indicating that this is not a minimum point. In addition, the negative derivative at x_b indicates that the minimum point will lie somewhere to the right of x_b (that is in the direction where the function decreases). Similarly, the positive derivative at x_c indicates both that this point is not a minimum and that the minimum lies to the left of x_c. These basic concepts form the foundation of gradient-based nonlinear optimization algorithms.

Nonlinear optimization problems can be categorized into three types. The simplest type is the unconstrained problem, which consists of the minimization of a single function without constraints. Two categories of constrained problems are possible: the linearly constrained nonlinear problem, where the objective function is nonlinear but the constraints are linear, and the nonlinearly constrained problem, where the constraints are nonlinear functions and the objective function may be linear or nonlinear.

Algorithm for Unconstrained Optimization

Gradient-based nonlinear optimization algorithms start from an initial estimate of the solution, q_0, and iteratively proceed to improve the so-

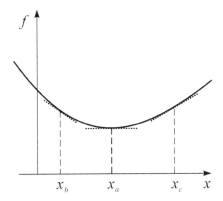

FIGURE 8.14 In gradient-based solution techniques, the function's gradient is used to determine a search direction for improving the solution.

lution, For most algorithms the choice of initial estimate is arbitrary. At each iteration, k, of the algorithm the solution point, $\mathbf{q}_k$, is updated. For minimization problems, each step of a gradient-based algorithm involves determining a direction along which the objective function will decrease (a descent direction), selecting a step length that minimizes the objective function along the selected direction, and taking the step. The procedure is demonstrated in the following text for minimization of a single-valued function, f, of multiple stresses, $\mathbf{q}_k$.

GRADIENT-BASED ALGORITHM

Step 0: Initialize iteration counter, k, and select starting stresses, $\mathbf{q}_0$.

$$k = 0, \qquad \mathbf{q}_k = \mathbf{q}_0 \qquad (8.40)$$

Step 1: Determine a direction, $\mathbf{d}_k$, to move.

$$\mathbf{d}_k = d\left(\mathbf{q}_k, \nabla_{\mathbf{q}_k} f, \nabla_{\mathbf{q}_k}^2 f, \mathbf{d}_{k-1}\right) \qquad (8.41)$$

Step 2: Find a minimum, by choice of step size α_k, of the objective function along that direction.

$$\text{minimize} \quad f(\mathbf{q}_k + \alpha_k \mathbf{d}_k) \qquad (8.42)$$

$$\text{such that} \quad 0 \leq \alpha_k \leq \alpha_{\max}$$

Step 3: Move to the new stress and increment the counter.

$$\mathbf{q}_{k+1} = \mathbf{q}_k + \alpha_k \mathbf{d}_k \qquad (8.43)$$

$$k = k + 1$$

Step 4: Test for convergence.

If convergence criteria satisfied, then stop; else go to step 1.

The direction found in step 1 is often determined as a function of some or all of the following: the current point, the objective value and its first and second derivatives at the current point, and the previous direction. For example, the steepest descent direction is defined as the gradient of the objective function times negative one. This direction is based on assuming a linear projection of the function and seeking the downhill direction on this linear approximation. The Newton direction presumes a quadratic approximation of the function and is determined using both the first and second derivatives of the function. Generally, the Newton direction will converge to the solution in fewer algorithmic iterations but is more expensive to compute at each iteration. Numerous other direction-finding methods are available.

The problem solved in step 2 is a one dimensional unconstrained line search in the variable α_k. A different value of α is determined at each iteration as indicated by the subscript. The step length determined in step 2 forms the basis for moving to the next solution as is computed in step 3.

Step 4 requires a test to determine whether convergence has occurred. For unconstrained minimization problems the derivative of the function will be zero at a local minimum. This property of the derivative suggests a convergence criterion, namely that the algorithm is deemed to have converged if the gradient of the objective function approaches zero.

Constrained Optimization

The steps of the gradient-based algorithm just described are the same for the constrained problem. However, both the means of determining a direction and the convergence criterion must be reconsidered when constraints are present.

It is possible that the constraints prevent the objective function from reaching its unconstrained minimum point. Hence, the optimal solution could be at a point where the gradient of the objective function is not zero. As such, the convergence criterion based solely on the objective function gradient is inadequate. Instead, an expanded set of conditions are employed to confirm convergence. These conditions involve derivatives of both the objective function and the constraints.

If the nonlinear problem is unconstrained, then any descent direction will improve the solution. In the presence of constraints a direction must be determined that both improves the solution and does not deviate substantially from satisfying the constraints. In the next two sections, means of determining directions in the presence of constraints are discussed.

Finding Directions on Linear Constraints

When the constraint functions are linear, the generally preferred methods use feasible directions. That is, we choose directions that guarantee that each successive solution point is feasible. This is easy to implement for linearly constrained problems. For problems with linear constraints, which can be written using a general constraint matrix $\mathbf{A}$ and a general right-hand side $\mathbf{b}$, a solution point is feasible if

$$\mathbf{Aq}_k = \mathbf{b} \tag{8.44}$$

To maintain feasibility, select a direction, $\mathbf{d}_k$, that satisfies

$$\mathbf{Ad}_k = \mathbf{0} \tag{8.45}$$

This requirement guarantees that if $\mathbf{q}_k$ satisfies (8.44), then $\mathbf{q}_{k+1}$ will also satisfy (8.44) because premultiplying (8.43) by $\mathbf{A}$ produces

$$\mathbf{Aq}_{k+1} = \mathbf{Aq}_k + \alpha_k \mathbf{Ad}_k \tag{8.46}$$

and substituting (8.44) and (8.45) into (8.46) results in the desired condition of feasibility for q_{k+1}. Determining a direction, d_k, that satisfies (8.45) can be accomplished by defining a vector space consisting of all vectors orthogonal to the rows of the constraint matrix A. A feasible direction is then any vector from this space.

Finding Directions on Nonlinear Constraints

Nonlinearly constrained optimization problems can be difficult because there may exist no linear directions that maintain feasibility on a nonlinear constraint set. Algorithms that solve this class of problems must allow the solution vector q_k to be infeasible at any particular iteration while progress is made on the objective function. Ultimately, the solution must be feasible, therefore the algorithm must trade off progress on minimizing the objective function with wandering too far from feasibility.

An approach widely taken with nonlinearly constrained problems is to solve a sequence of intermediate optimization problems (subproblems) that are constrained by linear approximations to the original nonlinear constraint functions. Thus, each subproblem is a linearly constrained problem, which can be solved using the well-developed methods for this class of problems. The sequence of subproblems converges to the solution of the original problem. The sequential linear programming algorithm presented in Section 8.3.1 is a form of this approach.

Another approach is to incorporate the nonlinear constraints directly into a modified objective function using a penalty method. With this approach the optimization problem (8.2) is reformulated as the unconstrained problem

$$\text{minimize} \quad F(q) = f(q_1, q_2, \ldots, q_n, h_1, h_2, \ldots, h_l) \tag{8.47}$$

$$+ M \sum_{k=1}^{m} \max\big(g_k(q_1, q_2, \ldots, q_n, h_1, h_2, \ldots, h_l), 0\big)$$

where M is a penalty parameter. Note that if all constraints are satisfied, the second term will be zero. If M is selected sufficiently large then deviations from the constraints will produce a large cost to the objective. Hence, a solution will be sought that drives the second term in the objective function to zero. Because (8.47) is an unconstrained problem, methods introduced earlier in this section can be used to solve it.

Limitations of Gradient-Based Methods

Gradient-based nonlinear optimization algorithms for minimizing an objective function, either constrained or unconstrained, are designed to start from an arbitrary point and move, through successive iterations, to a point where the convergence criteria are met. Gradient-based methods are designed to locate a local minimum of the problem. When using such an algorithm to solve a problem with a convex objective function on a convex feasible region, the local minimum thus found is also a global minimum. However, if the objective function or feasible region is not convex, then there is no guarantee that the local solution found is a global solution. For example, because of the apparent nonconvexity of the unconfined problem, gradient-based algorithms will terminate at a solution that may not be the global solution. In this circumstance the starting point used by the optimization algorithm is critical to determining which local solution will be found for a nonconvex problem.

This limitation may not be significant if there exist only a small number of local solutions. One way to test for alternative local minima is to initiate the algorithm from a series of different starting points. If the solutions found from alternative starting points are the same, then the analyst can have some assurance, although not be guaranteed, that the global optimal solution has been found.

8.3.3 NONGRADIENT SOLUTION METHODS

A number of algorithms have been developed that attempt to address the difficulties that arise when solving nonconvex problems. These methods are sometimes referred to as global optimization methods because of their ability to avoid convergence to local minimum points. However, no proof exists that they will actually find the global minimum. Furthermore, these methods are computationally intensive, which may limit their use to smaller problems. Recent work has focused on combining search techniques in an effort to minimize the computational burden while taking advantage of specific properties of the intensive methods. The methods based on natural analogues (e.g., simulated annealing and genetic algorithms) do not require continuity in any of the functions in the optimization problem. However, these methods are computationally intensive and best applied to problems for which traditional methods fail, such as problems with discontinuous, noisy, or multimodal properties. References to applications of these methods to groundwater management problems are given at the end of this chapter.

Cutting Plane Methods

Cutting plane methods enclose the feasible region in an approximating polytope and the objective function is minimized (or maximized) over this relaxed region. The enclosing polytope is a linear approximation of the constraint set, with the approximation improving at each iteration with each new cut. If the objective function is convex, then each subproblem is also convex and can be solved by a traditional convex programming approach. The solution converges from the infeasible region toward the feasible region and the global optimum. The technique requires that the objective function and constraints are continuous and uses special rules to govern cutting plane selection for nonconvex problems.

Simulated Annealing

Simulated annealing algorithms are based on an analogy with the thermodynamic process of cooling solids. When a material is melted and allowed to cool slowly (anneal), a crystal lattice emerges that minimizes the energy of the system. Simulated annealing uses this analogy to search for "minimum energy" configurations of the decision variables, where energy is represented by the objective function value for a given solution. The transition rules from one state, or configuration, to the next are applied stochastically, so that occasionally an inferior solution will be chosen over a better configuration. The probability of this occurring is larger at early stages of the annealing process than at later stages, when the temperature of the system has cooled. The stochastic nature of the search process ensures that the algorithm has a means of escaping local optima.

Genetic Algorithms

Genetic algorithms (GAs) are based on the theory of Darwinian evolution and the survival of the fittest. Whereas most optimization techniques make transitions from one solution to another, GAs retain a set of solutions and transition rules are applied to transform one set of solutions into another set. The basic analogy of GAs is that a set of solutions at an iteration represents a population, each solution represents an individual within the population, and the objective function value of each solution represents that individual's fitness. A new population, or generation, arises from the previous generation through competition based on fitness. The transition rules are applied on a stochastic basis so that the fittest individuals have the highest probability of being selected to produce the next generation. After the selection process, genetic operators, such as crossover and mutation, are applied to complete the transition from one generation to the next. As with simulated annealing, the stochastic transition rules provide GAs with a means of escaping from local optima.

The decision variables for the management model are analogous to biological genes and are represented by coded structures. Structure coding can take various forms, including real number representation and binary coding. A point in the search space is represented by a collection of genes. The coded genes are juxtaposed to form an individual or chromosome.

For groundwater problems, GAs are flexible enough to accommodate explicit representation of the stress location problem. An example of a coded solution is shown in Figure 8.15. In this example, the direction of stress (injection or extraction) is controlled by the first bit, the well's x-coordinate location is the second gene, the y-coordinate is encoded in the third gene,

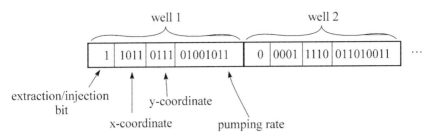

FIGURE 8.15   Example of decision variable coding in a genetic algorithm. In this case, four decision variables are associated with each candidate well: the direction of pumping, the x- and y-coordinates of the well, and the pumping rate.

and the last gene contains the stress rate. Minimum and maximum bounds for all of the decision variables are used to map the binary code to real numbers. Previous formulations presented in this book require a set of candidate locations whose locations are fixed. However, the decision variable structure depicted in Figure 8.15 allows the number of candidate stresses to be reduced by accommodating their locations explicitly as decision variables.

8.3.4 SOLVING NONLINEAR BINARY PROBLEMS

Problems that contain both nonlinear and binary elements arise from a number of sources. If any of the nonlinear features described before, such as unconfined flow or lift costs, are used in a problem with binary variables, then a nonlinear binary problem results. Certain formulations may also arise when these nonlinear features are not present but the nature of the binary formulation produces nonlinear products between decision variables, as demonstrated in Sections 8.2.4 and 8.2.5.

The combination of both nonlinear and binary elements produces a very difficult problem to solve. We first examine the potential for using the branch and bound method to solve the binary problem when nonlinearities are present. The basis for the branch and bound method is the successive separation of the feasible region into subregions. The assumption of the branch and bound method is that within each subregion the global optimum can be found. This global optimum is used as the basis for comparison between alternative branches. If the optimum in each region is not global, then this comparison is invalid and the method may not converge to the optimal solution.

If the initial feasible region is convex, then any linear subdivision of that region is also convex. If it is further assumed that the objective function is convex, then a global optimum can be assured in each of the subregions and the branch and bound approach will converge. However, if nonconvexity is present, the branch and bound algorithm may converge to a nonglobal minimum.

Other methods for addressing nonlinear binary problems have been proposed. For example, continuous functions that approximate the discontinuity, such as equation (8.24) or (8.25), have been used successfully.

The combinatorial techniques that have been described (e.g., simulated annealing and GAs) can incorporate fixed costs fairly easily. The presence of these discontinuities may require alterations in the control parameters for the algorithms but should not alter their ability to solve the problems. Although binary coding of the decision variables in GAs is not required, the coding provides a natural avenue for integrating binary variables into the problem. In fact, a binary variable can simply be mapped to a gene with only one bit. That bit will indicate whether the associated location is active or not in the same way that the binary variable does in the formulations presented in Chapter 7.

8.4 INTERPRETING SOLUTION RESULTS

As stated in Section 8.1.3, it is often difficult to determine whether a global optimum has indeed been found. Furthermore, nonlinearities preclude the readily available range analysis information obtained for linear programs (Section 5.2). Nevertheless, some local sensitivity information can be obtained at the optimal solution if the objective function and constraints are differentiable. Recall that in the linear case the shadow prices are obtained during solution by the simplex method and are interpreted as the marginal cost of a resource. A similar interpretation of nonlinear constraints can be obtained for the solution of continuous nonlinear problems. The first-order optimality condition states that at the optimum the gradient of the objective function is a linear combination of the constraint gradients. The linear coefficients are called Lagrange multipliers and are similar to shadow prices. For linear problems, the shadow prices are readily available from solution by the simplex method. Conversely, the Lagrange multipliers are readily given only if the solution is obtained by one of the Lagrange multiplier methods. Nonetheless, if the objective function and constraints are continuous and differentiable at the optimal solution, then the shadow prices can be determined. As for the linear case, the sensitivity information is local only to the neighborhood around the optimum.

Local sensitivity analysis can be performed with nonlinear problems using the perturbation approach presented in Chapter 5. However, nonlinear problems are generally difficult and time consuming to solve, so resolving the optimization problem for different parameter values is sometimes impractical.

8.5 NOTES AND REFERENCES

The properties of nonlinear functions and their relationship to optimization described in Section 8.1 can be found in numerous references for optimization and numerical analysis such as Luenberger (1984) and Johnson and Riess (1982).

Pumping costs related to lift, as described in Section 8.2.1, have been used by Maddock (1972) and by Willis and Finney (1985). An application of this cost function to the problem of minimizing energy costs has been reported by Rastogi (1989).

The combination of optimization with numerically simulated unconfined aquifers was first described by Aguado and Remson (1974). These authors exploit the fact that if the hydraulic conductivity is uniform and the system is modeled as steady state, the problem can be linearized by replacing the squared head term with a new linear variable. Maddock (1974) derives the nonlinear response function for the unconfined case in the form of an infinite series. Willis and Finney (1985) demonstrate several solution methods for the general transient unconfined optimization problem. Further applications to unconfined problems have been presented (Jones et al., 1987; Rizzo and Dougherty, 1996; Mansfield and Shoemaker, 1999). Colarullo et al. (1984) solved a hydraulic management problem for an unconfined aquifer that was modeled as confined and found that the error introduced by this approximation was small.

The use of concave nonlinear functions to approximate construction costs, described in Section 8.2.3, has been demonstrated in several groundwater applications. The exponential approximation has been successfully applied to groundwater problems by Karatzas and Pinder (1993). The polynomial approximation was presented by McKinney and Lin (1994, 1995) and Wang and Zheng (1998). McKinney and Lin (1994) also provide several examples of complex nonlinear objective functions that include additional capital and operational cost terms.

For a stream that may lose hydraulic connection with the aquifer, the determination of the proper flux by requiring that the product of fluxes be zero, presented in Section 8.2.4, equation 8.31, was used by Reichard (1987). Nonlinear boundary conditions that arise from drains are handled using a similar technique by Reichard (1995).

Further elaboration on the algorithms introduced in Section 8.3 can be found in Luenberger (1984), Bertsekas (1995), Goldberg (1989), and van Laarhoven and Aarts (1987). Numerical issues concerning the calculation of response matrix coefficients in unconfined groundwater flow management problems, as described in Section 8.3.1 are explored by Riefler and Ahlfeld (1996), who show that if the precision contained in the heads computed by the simulation is inadequate then there exists no perturbation value at which meaningful response coefficients can be determined. Another gradient-based algorithmic approach that has been extensively studied in groundwater applications is dynamic programming (Jones *et al.*, 1987; Makinde-Odusola and Marino, 1989; Andricevic, 1990; Chang *et al.*, 1992; Culver and Shoemaker, 1992). The nongradient solution methods introduced in Section 8.3.3 have been applied to a number of groundwater applications, especially those that include fixed costs. Algorithms such as simulated annealing (Dougherty and Marryott, 1991; Kuo *et al.*, 1992; Marryott *et al.*, 1993; Marryott, 1996; Wang and Zheng, 1998) and genetic algorithms (McKinney and Lin, 1994; Ritzel *et al.*, 1994; Huang and Mayer, 1997; Wang and Zheng, 1998) have been been found to be effective for these problems. Another non-gradient-based search technique, called tabu search, has been implemented by Zheng and Wang (1999). Tabu search is another search technique with a natural analogue, in this case the human memory process.

Appendix

Software for Groundwater Management

The CD-ROM attached to this book contains a complete software package that can be used to solve some of the problem types described in this book. The package is called MODOFC (MODflow Optimal Flow Control). MODOFC couples with the USGS MODFLOW simulation program and contains a full implementation of the simplex algorithm. MODOFC has the following capabilities:

Objective Function Components

- Linear pumping costs
- Installation costs

Constraints

- Upper and lower bounds on head
- Upper and lower bounds on individual stresses
- Upper and lower bounds on net pumping in each stress period
- Bounds on head difference (or gradient)
- Total injection pumping is a fraction of total extraction pumping
- Bounds on total number of stress locations to be used

Interfaces with MODFLOW Capabilities

- Incorporates MODFLOW96 version of MODFLOW
- Confined or unconfined units
- Wells screened in single or multiple layers
- Single or multiple stress periods

- Stresses active at a single rate in any combination of stress periods
- Stresses active at different rates in any combination of stress periods
- Head bounds, head difference bounds, or capture zones active at a single constraint value in any combination of stress periods
- Individual stress bounds active in each stress period
- Total stress bounds active in each stress period
- Specified injection fraction active in all stress periods

Solution Algorithms Used

- Simplex algorithm for linear problems
- Sequential linear programming for unconfined problems
- Branch and bound for mixed binary problems

Output Reported

- Optimal pumping rates and locations
- Binding constraints and shadow prices
- Range analysis
- Input and run-time error messages

Accuracy and Efficiency Features

- User selection of perturbation level and capability for variable perturbation level
- Dynamic dimensioning coupled with MODFLOW memory needs
- Response matrix retained for repeated solution
- Capture zone input as a series of line segments

The CD-ROM contains

1. Complete FORTRAN source code for MODOFC.
2. An executable compiled using the Lahey F77L-EM/32 FORTRAN 77 Compiler and suitable for running under DOS.
3. Documentation for MODOFC written by David Ahlfeld and R. Guy Riefler provided as a Microsoft Word 7.0 file.
4. Several sample problems that are described in the documentation
5. The HTML files from the MODOFC Web site, which contains additional information on groundwater management models and links to the ongoing MODOFC Web site maintained by the authors.
6. A README file that describes the directory structures and file characteristics in more detail.

The CD-ROM is readable from DOS and has been successfully accessed from Windows 95, Windows 98, and Windows NT.

The contents of this CD-ROM and other information on groundwater management can be obtained via the World Wide Web at http://www.ecs.umass.edu/modofc/

REFERENCES

Aguado, E., and I. Remson, 1974, Groundwater hydraulics in aquifer management, Journal of the Hydraulics Division, American Society of Civil Engineers, 100(HY1), pp. 103–118.

Aguado, E., and I. Remson, 1980, Groundwater management with fixed charges, Journal of Water Resources Planning and Management Division, ASCE, 106(2), pp. 375–382.

Aguado, E., I. Remson, M. F. Pikul, and W. A. Thomas, 1974, Optimal pumping for aquifer dewatering, Journal of the Hydraulics Division, ASCE, 100(7), pp. 869–877.

Ahlfeld, D. P., 1990, Two-stage groundwater remediation design, Journal of Water Resources Planning and Management, ASCE, 116(4), pp. 517–529.

Ahlfeld, D. P., 1998, Interpretation of the dual program for optimal groundwater hydraulic control, Journal of the American Water Resources Association, 34(1), pp. 195–206.

Ahlfeld, D. P., and M. Heidari, 1994, Applications of optimal hydraulic control to ground-water systems, Journal of Water Resources Planning and Management, ASCE, 120(3), pp. 350–365.

Ahlfeld, D. P., R. H. Page, and G. F. Pinder, 1995, Optimal ground-water remediation methods applied to a Superfund site: From formulation to implementation, Ground Water, 33(1), pp.58–70.

Ahlfeld, D. P., and C. S. Sawyer, 1990, Well location in capture zone design using simulation and optimization techniques, Ground Water, 28(4), pp. 507–512.

Alley, W. M., E. Aguado, and I. Remson, 1976, Aquifer management under transient and steady-state conditions, Water Resources Bulletin, 12(5), pp. 963–972.

Anderson, M. P., and W. W. Woessner, 1992, Applied Groundwater Modeling: Simulation of Flow and Advective Transport, Academic Press, San Diego, CA.

Andricevic, R., 1990, A real-time approach to managment and monitoring of groundwater hydraulics, Water Resources Research, 26(11), pp. 2747–2755.

Atwood, D. V. , and S. M. Gorelick, 1985, Hydraulic gradient control for ground-water contaminant removal, Journal of Hydrology, 76, pp. 85–106.

Barlow, P. M., B. J. Wagner, and K. Belitz, 1996, Pumping strategies for management of a shallow water table: The value of the simulation–optimization approach, Ground Water, 34(2), pp. 305–317.

Bear, J., 1972, Dynamics of Fluids in Porous Media, American Elsevier Publishing Company, New York.

Bear, J., 1979, Hydraulics of Groundwater, McGraw-Hill, New York.

Bertsekas, D. P., 1995, Nonlinear Programming, Athena Scientific, Belmont, MA.

Bradley, S. P., A. C. Hax, and T. L. Magnanti, 1977, Applied Mathematical Programming, Addison-Wesley Publishing Company, Reading, MA.

Celia, M. A., and W. G. Gray, 1992, Numerical Methods for Differential Equations, Fundamental Concepts for Scientific and Engineering Applications, Prentice Hall, Englewood Cliffs, NJ.

Chan, N., 1993, Robustness of the multiple realization method for stochastic hydraulic aquifer management, Water Resources Research, 29(9), pp. 3159–3167.

Chan, N., 1994, Partial infeasibility method for change-constrained aquifer management, Journal of Water Resources Planning and Management, 120 (1), pp. 70–89.

Chang, L.-C., C. A. Shoemaker, and P. L.-F. Liu, 1992, Optimal time-varying pumping rates for groundwater remediation: Application of a constrained optimal control algorithm, Water Resources Research, 28(12), pp. 3157–3173.

Charnes, A., and W. W. Cooper, 1959, Chance-constrained programming, Management Science, 6(1), pp. 73–79.

Collarulo, S. J., M. Heidari, and T. Maddock III, 1984, Identification of an optimal groundwater management strategy in a contaminated aquifer, Water Resources Bulletin, 20(5), pp. 747–760.

Collarulo, S. J., M. Heidari, and T. Maddock III, 1985, Demonstrative model for identifying ground-water-management options in a contaminated aquifer, Groundwater Series 8, Kansas Geological Survey, Lawrence, KS.

Culver, T. B., and C. A. Shoemaker, 1992, Dynamic optimal control for groundwater remediation with flexible management periods, Water Resources Research, 28(3), pp. 629–641.

Danskin, W. R., and J. R. Freckleton, 1992, Ground-water-flow modeling and optimization techniques applied to high ground-water problems in San Bernardino, California, in Selected Papers in the Hydrologic Sciences, S. Subitzky (ed.), U.S. Geol. Surv. Water-Supply Paper 2340, pp. 165–177.

Danskin, W. R., and S. M. Gorelick, 1985, Policy evaluation tool: Management of a multi-aquifer system using controlled stream recharge, Water Resources Research, 21(11), pp. 1731–1747.

Dantzig, G. B., 1963, Linear Programming and Extensions, Princeton University Press, Princeton, NJ.

de Marsily, G., 1986, Quantitative Hydrogeology, Academic Press, San Diego, CA.

Deninger, R. A., 1970, Systems analysis of water supply systems, Water Resources Bulletin, 6(4), pp. 573–579.

Domenico, P. A., and F. W. Schwartz, 1997, Physical and Chemical Hydrogeology, John Wiley & Sons, New York.

Dougherty, D. E., and R. A. Marryott, 1991, Optimal groundwater management, 1. Simulated annealing, Water Resources Research, 27(10), pp. 2493–2508.

Duckstein, L., W. Treichel, and S. El Magnouni, 1994, Ranking ground-water management alternatives by multicriterion analysis, Journal of Water Resources Planning and Management, 120(4), pp. 546–563.

El Magnouni, S., and W. Treichel, 1994, A multicriterion approach to groundwater management, Water Resources Research, 30(6), pp. 1881–1895.

Emch, P. G., and W. W-G. Yeh, 1998, Management model for conjunctive use of coastal surface water and groundwater, Journal of Water Resources Planning and Management, 124(3), pp. 129–138.

Finney, B.A., Samsuhadi, and R. Willis, 1992, Quasi three-dimensional optimization model of Jakarta basin, Journal of Water Resources Planning and Management, 118(1), pp.18–31.

Francko, D. A., and R. G. Wetzel, 1983, To Quench our Thirst: The Present and Future Status of Freshwater Resources of the United States, The University of Michigan Press, Ann Arbor, MI.

Freeze, R. A., and J. A. Cherry, 1979, Groundwater, Prentice-Hall, Englewood Cliffs, NJ.

Galeati, G., and G. Gambolati, 1988, Optimal dewatering schemes in the foundation design of an electronuclear plant, Water Resources Research, 24(4), pp. 541–552.

Garfinkel, R. S., and G. L. Nemhauser, 1972, Integer Programming, Wiley, New York.

Gharbi, A., and R. C. Peralta, 1994, Integrated embedding optimization applied to Salt Lake valley aquifers, Water Resources Research, 30(3), pp. 817–832.

Gill, P. E., W. Murray, and M. H. Wright, 1981, Practical Optimization, Academic Press, New York.

Gleick, P. H., ed., 1993, Water in Crises: A Guide to the World's Fresh Water Resources, Oxford University Press, New York.

Goldberg, D. E., 1989, Genetic Algorithms in Search, Optimization, and Machine Learning, Addison-Wesley Publishing Company, Reading, MA.

Gorelick, S. M., 1982, A model for managing sources of groundwater pollution, Water Resources Research, 18(4), pp. 773–781.

Gorelick, S. M., 1983, A review of distributed parameter ground-water management modeling methods, Water Resources Research, 19(2), pp. 305–319.

Gorelick, S. M., and I. Remson, 1982, Optimal location and management of waste disposal facilities affecting ground water quality, Water Resource Bulletin, 18(1), pp. 43–51.

Gorelick, S. M., C. I. Voss, P. E. Gill, W. Murray, M. A. Saunders, and M. H. Wright, 1984, Aquifer reclamation design: The use of contaminant transport simulation combined with nonlinear programming, Water Resources Research, 20(4), pp. 415–427.

Gorelick, S. M., R. A. Freeze, D. Donohue, and J. F. Keely, 1993, Groundwater Contamination: Optimal Capture and Containment, Lewis Publishers, Chelsea, MI.

Haggerty, R., and S. M. Gorelick, 1994, Design of multiple contaminant remediation: Sensitivity to rate-limited mass transfer, Water Resources Research, 30(2), pp. 435–446.

Hallaji, K., and H. Yazicigil, 1996, Optimal management of a coastal aquifer in southern Turkey, Journal of Water Resources Planning and Management, 122(4), pp. 233–244.

Harbaugh, A. W., and M. G. McDonald, 1996, User's documentation for MODFLOW-96, an update to the U.S. Geological Survey modular finite-difference ground-water flow model: U.S. Geological Survey Open-File Report 96-485, 56 pp.

Heidari, M., J. Sadeghipour, and O. Drici, 1987, Velocity control as a tool for optimal plume management in the Equus Beds aquifer, Kansas, Water Resources Bulletin, 23(2), pp. 325–336.

Hillier, F. S., and G. J. Lieberman, 1995, Introduction to Operations Research, McGraw-Hill, New York.

Hinrichsen, D., B. Robey, and U. D. Upadhyay, 1997, Solutions for a Water-Short World, Population Reports, Series M, No. 14, Johns Hopkins School of Public Health, Population Information Program, Baltimore, MD.

Huang, C., and A. S. Mayer, 1997, Pump-and-treat optimization using well locations and pumping rates as decision variables, Water Resources Research, 33(5), pp. 1001–1012.

Huyakorn, P. S., and G. F. Pinder, 1983, Computational Methods in Subsurface Flow, Academic Press, New York.

Johnson, L. W., and R. D. Riess, 1982, Numerical Analysis, Second Edition, Addison Wesley Publishing Company, Reading, MA.

Jones, L., R. Willis, and W. W.-G. Yeh, 1987, Optimal control of nonlinear groundwater hydraulics using differential dynamic programming, Water Resources Research, 23(11), pp. 2097–2106.

Karatzas, G. P., and G. F. Pinder, 1993, Groundwater management using numerical simulation and the outer approximation method for global optimization, Water Resources Research, 29(10), pp. 3371–3378.

Kuo, C.-H., A. N. Michel, and W. G. Gray, 1992, Design of optimal pump-and-treat strategies for contaminated groundwater remediation using simulated annealing algorithm, Advances in Water Resources, 15, pp. 95–105.

Lall, U., and Y. W. H. Lin, 1991, A groundwater management model for Salt Lake County, Utah with some water rights and water quality considerations, Journal of Hydrology, 123, pp. 367–393.

Lall, U., and M. D. Santini, 1989, An optimization model for unconfined stratified aquifer systems, Journal of Hydrology, 111, pp. 145–162.

Lee, A. S., and J. S. Aronofsky, 1958, A linear programming model for scheduling crude oil production, JPT Journal of Petroleum Technology, 213, pp. 51–54.

Lefkoff, L. J., and S. M. Gorelick, 1985, Rapid removal of a ground-water contaminant plume, in Groundwater Contamination and Reclamation, American Water Resources Association, pp. 125–131.

Luenberger, D. G., 1984, Linear and Nonlinear Programming, second edition, Addison-Wesley Publishing Company, Reading, MA.

Maddock, T., III, 1972, Algebraic technological function from a simulation model, Water Resources Research, 8(1), pp. 129–134.

Maddock, T., III, 1973, Management model as a tool for studying the worth of data, Water Resources Research, 9(3), pp. 270–280.

Maddock, T., III, 1974, Nonlinear technological functions for aquifers whose transmissivities vary with drawdown, Water Resources Research, 10(4), pp. 877–881.

Makinde-Odusola, B. A., and M. A. Marino, 1989, Optimal control of groundwater by the feedback method of control, Water Resources Research, 25(6), pp. 1341–1352.

Mansfield, C. M., and C. A. Shoemaker, 1999, Optimal remediation of unconfined aquifers: Numerical applications and derivative calculations, Water Resources Research, 35(5), pp. 1455–1469.

Marryott, R. A., 1996, Remediation design using multiple control technologies, Ground Water, 34(3), pp. 425–433.

Marryott, R. A., D. E. Dougherty, and R. L. Stollar, 1993, Optimal groundwater management 2. Application of simulated annealing to a field-scale contamination site, Water Resources Research, 29(4), pp. 847–860.

Mays, L. W., 1997, Optimal Control of Hydrosystems, Marcel Dekker, New York.

Mays, L. W., and Y.-K. Tung, 1992, Hydrosystems Engineering and Management, McGraw-Hill, New York.

McKinney, D. C., and M.-D. Lin, 1994, Genetic algorithm solution of groundwater management models, Water Resources Research, 30(6), pp. 1897–1906.

McKinney, D. C., and M.-D. Lin, 1995, Approximate mixed-integer nonlinear programming methods for optimal aquifer remediation design, Water Resources Research, 31(3), pp. 731–740.

Minsker, B. S., and C. A. Shoemaker, 1996, Differentiating a finite element biodegradation simulation model for optimal control, Water Resources Research, 32(1), pp. 187–192.

Misirli, F., and H. Yazicigil, 1997, Optimal ground-water pollution plume containment with fixed charges, Journal of Water Resources Planning and Management, 123(1), pp. 2–14.

Molz, F. J., and L. C. Bell, 1977, Head gradient control in aquifers used for fluid storage, Water Resources Research, 13(4), pp. 795–798.

Morgan, D. R., J. W. Eheart, and A. J. Valoocchi, 1993, Aquifer remediation design under uncertainty using a new chance constrained programming technique, Water Resources Research, 29(3), pp. 551–561.

Murtagh, B. A., 1981, Advanced Linear Programming: Computation and Practice, McGraw-Hill, New York.

Nazareth, J. L., 1987, Computer Solution of Linear Programs, Oxford University Press, New York.

Nemhauser, G. L., and L. A. Wolsey, 1988, Integer and Combinatorial Optimization, Wiley, New York.

Nishikawa, T., 1998, Water resources optimization model for Santa Barbara, California, Journal of Water Resources Planning and Management, 124(5), pp. 252–263.

Peralta, R. C., and P. J. Killian, 1985, Optimal regional potentiometric surface design: Least-cost water supply/sustained groundwater yield, Transactions of the American Society of Agricultural Engineers, 28(4), pp. 1098–1107.

Peralta, R. C., H. Azarmnia, and S. Takahashi, 1991, Embedding and response matrix techniques for maximizing steady-state ground-water extraction: Computational comparison, Ground Water, 29(3), pp. 357–364.

Peralta, R. C., R. R. A. Cantiller, and J. E. Terry, 1995, Optimal large-scale conjunctive water-use planning: Case study, Journal of Water Resources Planning and Management, 121(6), pp. 471–478.

Rastogi, A. K., 1989, Optimal pumping policy and groundwater balance for the Blue Lake aquifer, California, involving nonlinear groundwater hydraulics, Journal of Hydrology, 111(3), pp. 177–194.

Ratzlaff, S. A., M. M. Aral, and F. Al Khayyal, 1992, Optimal design of groundwater capture systems using segmental velocity–direction constraints, Ground Water, 30(4), pp. 607–612.

Reichard, E. G., 1987, Hydrologic influences on the potential benefits of basinwide groundwater management, Water Resources Research, 23(1), pp. 77–91.

Reichard, E. G., 1995, Groundwater–surface water management with stochastic surface water supplies: A simulation optimization approach, Water Resources Research, 31(11), pp. 2845–2865.

Riefler, R. G. and D. P. Ahlfeld, 1996, The impact of numerical precision on the solution of confined and unconfined optimal hydraulic control problems, Hazardous Waste & Hazardous Materials, 13, pp. 167–176.

Ritzel, B. J., J. W. Eheart, and S. Ranjithan, 1994, Using genetic algorithms to solve a multiple objective groundwater pollution containment problem, Water Resources Research, 30(5), pp. 1589–1603.

Rizzo, D. M. and D. E. Dougherty, 1996, Design optimization for multiple management period groundwater remediation, Water Resources Research, 32(8), pp. 2549–2561.

Sawyer, C. S., and Y.-F. Lin, 1998, Mixed-integer chance-constrained models for groundwater remediation, Journal of Water Resources Planning and Management, 124(5), pp. 285–294.

Sawyer, C. S., D. P. Ahlfeld, and A. J. King, 1995, Groundwater remediation design using a three-dimensional simulation model and mixed-integer programming, Water Resources Research, 31(5), pp. 1373–1385.

Shafike, N. G., L. Duckstein, and T. Maddock III, 1992, Multicriterion analysis of groundwater contamination management, Water Resources Bulletin, 28(1), pp. 33–43.

Shamir, U., J. Bear, and A. Gamliel, 1984, Optimal annual operation of a coastal aquifer, Water Resources Research, 20(4), pp. 435–444.

Solow, D., 1984, Linear Programming. An Introduction to Finite Improvement Algorithms, Elsevier Science Publishing, New York.

Sun, Y.-H., M. W. Davert, and W. W.-G. Yeh, 1996, Soil vapor extraction system design by combinatorial optimization, Water Resources Research, 32(6), pp. 1863–1873.

Sykes, J. F., J. L. Wilson, and R. W. Andrews, 1985, Sensitivity analysis for steady-state groundwater flow using adjoint operators, Water Resources Research, 21(3), p. 359–371.

Thompson, S. A., 1999, Water Use, Management, and Planning in the United States, Academic Press, San Diego, CA.

Tiedeman, C., and S. M. Gorelick, 1993, Analysis of uncertainty in optimal groundwater contaminant capture design, Water Resources Research, 29(7), pp. 2139–2153.

Tung, Y.-K., 1986, Groundwater management by chance-constrained model, Journal of Water Resources Planning and Management, 112(1), pp. 1–19.

Van der Leeden, F., F. L. Troise, and D. K. Todd, 1990, The Water Encyclopedia, second edition, Lewis Publishers, Chelsea, MI.

van Laarhoven, P. J. M., and E. H. L. Aarts, 1987, Simulated Annealing: Theory and Applications, Kluwer Academic, Boston, MA.

Verdon C A, 1995, Measure of robustness for a linear groundwater optimization problem, M.S. Thesis, University of Connecticut, Storrs, CT.

Veselov, V, V, V. M. Miilas, and V. P. Stepanenko, 1992, Questions of Modeling and Optimization of Hydrogeological Systems, Gylym, Alma-Ata, Kazakhstan (in Russian).

Wagner, B. J., 1995, Recent advances in simulation-optimization groundwater management modeling, Reviews of Geophysics, Supplement, 33, pp. 1021–1028.

Wagner, B. J., and S. M. Gorelick, 1987, Optimal groundwater quality management under parameter uncertainty, Water Resources Research, 23(7), pp. 1162–1174.

Wagner, B. J., and S. M. Gorelick, 1989, Reliable aquifer remediation in the presence of spatially variable hydraulic conductivity: From data to design, Water Resources Research, 25(10), pp. 2211–2225.

Wagner, J. M., U. Shamir, and H. R. Nemati, 1992, Groundwater quality management under uncertainty: Stochastic programming approaches and the value of information, Water Resources Research, 28(5), pp. 1233–1246.

Wang, M., and C. Zheng, 1998, Ground water management optimization using genetic algorithms and simulated annealing: Formulation and comparison, Journal of the American Water Resources Association, 34(3), pp. 519–530.

Watkins, D. W., and D. C. McKinney, 1997, Finding robust solutions to water resources problems, Journal of Water Resources Planning and Management, 123(1), pp. 49–58.

Willis, R., and B. A. Finney, 1985, Optimal control of nonlinear groundwater hydraulic: Theoretical development and numerical experiments, Water Resources Research, 21(10), pp. 1476–1482.

Willis, R., and W. W.-G. Yeh, 1987, Groundwater Systems Planning and Management, Prentice-Hall, Englewood Cliffs, NJ.

Yeh, W. W.-G., 1992, Systems analysis in ground-water planning and management, Journal of Water Resources Planning and Management, 118(3), pp. 224–237.

Zhen, C, and J. G. Uber, 1996, Reliability of remediation designs in presence of modeling error, Journal of Water Resources Planning and Management, 122(4), pp. 253–261.

Zheng, C., and P. P. Wang, 1999, An integrated global and local optimization approach for remediation system design, Water Resources Research, 35(1), pp. 137–148.

INDEX

DATE DUE

GAYLORD No. 2333   PRINTED IN U.S.A.